PREFACE

In this edition of my *Logic*, the text has been revised throughout, several passages have been rewritten, and some sections added. The chief alterations and additions occur in cc. i., v., ix., xiii., xvi., xvii., xx.

The work may be considered, on the whole, as attached to the school of Mill; to whose *System of Logic*, and to Bain's *Logic*, it is deeply indebted. Amongst the works of living writers, the *Empirical Logic* of Dr. Venn and the *Formal Logic* of Dr. Keynes have given me most assistance. To some others acknowledgments have been made as occasion arose.

For the further study of contemporary opinion, accessible in English, one may turn to such works as Mr. Bradley's *Principles of Logic*, Dr. Bosanquet's *Logic; or the Morphology of Knowledge*, Prof. Hobhouse's *Theory of Knowledge*, Jevon's *Principles of Science*, and Sigwart's *Logic*. Ueberweg's *Logic, and History of Logical Doctrine* is invaluable for the history of our subject. The attitude toward Logic of the Pragmatists or Humanists may best be studied in Dr. Schiller's *Formal Logic*, and in Mr. Alfred Sidgwick's *Process of Argument* and recent *Elementary Logic*. The second part of this last work, on the "Risks of Reasoning," gives an admirably succinct account of their position. I agree with the Humanists that, in all argument, the important thing to attend to is the meaning, and that the most serious difficulties of reasoning occur in dealing with the matter reasoned about; but I find that a pure science of relation has a necessary place in the system of knowledge, and that the formulæ known as laws of contradiction, syllogism and causation are useful guides in the framing and testing of arguments and experiments concerning matters of fact. Incisive criticism of traditionary doctrines, with some remarkable reconstructions, may be read in Dr. Mercier's *New Logic*.

In preparing successive editions of this book, I have profited by the comments of my friends: Mr. Thomas Whittaker, Prof. Claude Thompson, Dr. Armitage Smith, Mr. Alfred Sidgwick, Dr. Schiller, Prof. Spearman, and Prof. Sully, have made important suggestions; and I might have profited more by them, if the frame of my book, or my principles, had been more elastic.

As to the present edition, useful criticisms have been received from Mr. S.C. Dutt, of Cotton College, Assam, and from Prof. M.A. Roy, of Midnapore; and, especially, I must heartily thank my colleague, Dr. Wolf, for communications that have left their impress upon nearly every chapter.

Carveth Read.

London,
August, 1914

Contents

PREFACE 3

LOGIC 7

CHAPTER I 7

    INTRODUCTORY 7

CHAPTER II 10

    GENERAL ANALYSIS OF PROPOSITIONS 10

CHAPTER III 13

    OF TERMS AND THEIR DENOTATION 13

CHAPTER IV 15

    THE CONNOTATION OF TERMS 15

CHAPTER V 19

    THE CLASSIFICATION OF PROPOSITIONS 19

CHAPTER VI 24

    CONDITIONS OF IMMEDIATE INFERENCE 24

CHAPTER VII 26

    IMMEDIATE INFERENCES 26

CHAPTER VIII 31

    ORDER OF TERMS, EULER'S DIAGRAMS, LOGICAL EQUATIONS, EXISTENTIAL IMPORT OF PROPOSITIONS 31

CHAPTER IX 35

    FORMAL CONDITIONS OF MEDIATE INFERENCE 35

CHAPTER X 38

    CATEGORICAL SYLLOGISMS 38

CHAPTER XI 46

ABBREVIATED AND COMPOUND ARGUMENTS 46

CHAPTER XII 49

CONDITIONAL SYLLOGISMS 49

CHAPTER XIII 53

TRANSITION TO INDUCTION 53

CHAPTER XIV 57

CAUSATION 57

CHAPTER XV 61

INDUCTIVE METHOD 61

A. Qualitative Determination 64

B. Quantitative Determination 64

CHAPTER XVI 65

THE CANONS OF DIRECT INDUCTION 65

(I) The Canon of Agreement. 65

§ 2. The Canon of the Joint Method of Agreement in Presence and in Absence. 66

§ 3. The Canon of Difference. 67

§ 4. The Canon Of Concomitant Variations. 69

§ 5. The Canon Of Residues. 72

CHAPTER XVII 73

COMBINATION OF INDUCTION WITH DEDUCTION 73

CHAPTER XVIII 80

HYPOTHESES 80

CHAPTER XIX 86

LAWS CLASSIFIED; EXPLANATION; CO-EXISTENCE; ANALOGY 86

CHAPTER XX 91

PROBABILITY 91

CHAPTER XXI	97
DIVISION AND CLASSIFICATION	97
CHAPTER XXII	103
NOMENCLATURE, DEFINITION, PREDICABLES	103
CHAPTER XXIII	109
DEFINITION OF COMMON TERMS	109
CHAPTER XXIV	113
FALLACIES	113

LOGIC

CHAPTER I
INTRODUCTORY

§ 1. Logic is the science that explains what conditions must be fulfilled in order that a proposition may be proved, if it admits of proof. Not, indeed, every such proposition; for as to those that declare the equality or inequality of numbers or other magnitudes, to explain the conditions of their proof belongs to Mathematics: they are said to be *quantitative*. But as to all other propositions, called *qualitative*, like most of those that we meet with in conversation, in literature, in politics, and even in sciences so far as they are not treated mathematically (say, Botany and Psychology); propositions that merely tell us that something happens (as that *salt dissolves in water*), or that something has a certain property (as that *ice is cold*): as to these, it belongs to Logic to show how we may judge whether they are true, or false, or doubtful. When propositions are expressed with the universality and definiteness that belong to scientific statements, they are called laws; and laws, so far as they are not laws of quantity, are tested by the principles of Logic, if they at all admit of proof.

But it is plain that the process of proving cannot go on for ever; something must be taken for granted; and this is usually considered to be the case (1) with particular facts that can only be perceived and observed, and (2) with those highest laws that are called 'axioms' or 'first principles,' of which we can only say that we know of no exceptions to them, that we cannot help believing them, and that they are indispensable to science and to consistent thought. Logic, then, may be briefly defined as the science of proof with respect to *qualitative* laws and propositions, except those that are axiomatic.

§ 2. Proof may be of different degrees or stages of completeness. Absolute proof would require that a proposition should be shown to agree with all experience and with the systematic explanation of experience, to be a necessary part of an all-embracing and self-consistent philosophy or theory of the universe; but as no one hitherto has been able to frame such a philosophy, we must at present put up with something less than absolute proof. Logic, assuming certain principles to be true of experience, or at least to be conditions of consistent discourse, distinguishes the kinds of propositions that can be shown to agree with these principles, and explains by what means the agreement can best be exhibited. Such principles are those of Contradiction (chap. vi.), the Syllogism (chap. ix.), Causation (chap. xiv.), and Probabilities (chap. xx.). To bring a proposition or an argument under them, or to show that it agrees with them, is logical proof.

The extent to which proof is requisite, again, depends upon the present purpose: if our aim be general truth for its own sake, a systematic investigation is necessary; but if our object be merely to remove some occasional doubt that has occurred to ourselves or to others, it may be enough to appeal to any evidence that is admitted or not questioned. Thus, if a man doubts that *some acids are compounds of oxygen*, but grants that *some compounds of oxygen are acids*, he may agree to the former proposition when you point out that it has the same meaning as the latter, differing from it only in the order of the words. This is called proof by immediate inference.

Again, suppose that a man holds in his hand a piece of yellow metal, which he asserts to be copper, and that we doubt this, perhaps suggesting that it is really gold. Then he may propose to dip it in vinegar; whilst we agree that, if it then turns green, it is copper and not gold. On trying this experiment the metal does turn green; so that we may put his argument in this way:—

Whatever yellow metal turns green in vinegar is copper; This yellow metal turns green in vinegar; Therefore, this yellow metal is copper.

Such an argument is called proof by mediate inference; because one cannot see directly that the yellow metal is copper; but it is admitted that any yellow metal is copper that turns green in vinegar, and we are shown that this yellow metal has that property.

Now, however, it may occur to us, that the liquid in which the metal was dipped was not vinegar, or not pure vinegar, and that the greenness was due to the impurity. Our friend must thereupon show by some means that the vinegar was pure; and then his argument will be that, since nothing but the vinegar came in contact with the metal, the greenness was due to the vinegar; or, in other words, that contact with that vinegar was the cause of the metal turning green.

Still, on second thoughts, we may suspect that we had formerly conceded too much; we may reflect that, although it had often been shown that copper turned green in vinegar, whilst gold did not, yet the same might not always happen. May it not be, we might ask, that just at this moment, and perhaps always for the future gold turns, and will turn green in vinegar, whilst copper does not and never will again? He will probably reply that this is to doubt the uniformity of causation: he may hope that we are not serious: he may point out to us that in every action of our life we take such uniformity for granted. But he will be obliged to admit that, whatever he may say to induce us to assent to the principle of Nature's uniformity, his arguments will not amount to logical proof, because every argument in some way assumes that principle. He has come, in fact, to the limits of Logic. Just as Euclid does not try to prove that 'two magnitudes equal to the same third are equal to one another,' so the Logician (as such) does not attempt to prove the uniformity of causation and the other principles of his science.

Even when our purpose is to ascertain some general truth, the results of systematic inquiry may have various degrees of certainty. If Logic were confined to strict demonstration, it would cover a narrow field. The greater part of our conclusions can only be more or less probable. It may, indeed, be maintained, not unreasonably, that no judgments concerning matters of fact can be more than probable. Some say that all scientific results should be considered as giving the average of cases, from which deviations are to be expected. Many matters can only be treated statistically and by the methods of Probability. Our ordinary beliefs are adopted without any methodical examination. But it is the aim, and it is characteristic, of a rational mind to distinguish degrees of certainty, and to hold each judgment with the degree of confidence that it deserves, considering the evidence for and against it. It takes a long time, and much self-discipline, to make some progress toward rationality; for there are many causes of belief that are not good grounds for it—have no value as evidence. Evidence consists of (1) observation; (2) reasoning checked by observation and by logical principles; (3) memory—often inaccurate; (4) testimony—often untrustworthy, but indispensable, since all we learn from books or from other men is taken on testimony; (5) the agreement of all our results. On the other hand, belief is caused by many influences that are not evidence at all: such are (1) desire, which makes us believe in whatever serves our purpose; fear and suspicion, which (paradoxically) make us believe in whatever seems dangerous; (2) habit, which resists whatever disturbs our prejudices; (3) vanity, which delights to think oneself always right and consistent and disowns fallibility; (4)

imitativeness, suggestibility, fashion, which carry us along with the crowd. All these, and nobler things, such as love and fidelity, fix our attention upon whatever seems to support our prejudices, and prevent our attending to any facts or arguments that threaten to overthrow them.

§ 3. Two departments of Logic are usually recognised, Deduction and Induction; that is, to describe them briefly, proof from principles, and proof from facts. Classification is sometimes made a third department; sometimes its topics are distributed amongst those of the former two. In the present work the order adopted is, Deduction in chaps. ii. to xiii.; Induction in chaps. xiii. to xx.; and, lastly, Classification. But such divisions do not represent fundamentally distinct and opposed aspects of the science. For although, in discussing any question with an opponent who makes admissions, it may be possible to combat his views with merely deductive arguments based upon his admissions; yet in any question of general truth, Induction and Deduction are mutually dependent and imply one another.

This may be seen in one of the above examples. It was argued that a certain metal must be copper, because every metal is copper that turns green when dipped in vinegar. So far the proof appealed to a general proposition, and was deductive. But when we ask how the general proposition is known to be true, experiments or facts must be alleged; and this is inductive evidence. Deduction then depends on Induction. But if we ask, again, how any number of past experiments can prove a general proposition, which must be good for the future as well as for the past, the uniformity of causation is invoked; that is, appeal is made to a principle, and that again is deductive proof. Induction then depends upon Deduction.

We may put it in this way: Deduction depends on Induction, if general propositions are only known to us through the facts: Induction depends on Deduction, because one fact can never prove another, except so far as what is true of the one is true of the other and of any other of the same kind; and because, to exhibit this resemblance of the facts, it must be stated in a general proposition.

§ 4. The use of Logic is often disputed: those who have not studied it, often feel confident of their ability to do without it; those who have studied it, are sometimes disgusted with what they consider to be its superficial analysis of the grounds of evidence, or needless technicality in the discussion of details. As to those who, not having studied Logic, yet despise it, there will be time enough to discuss its utility with them, when they know something about it; and as for those who, having studied it, turn away in disgust, whether they are justified every man must judge for himself, when he has attained to equal proficiency in the subject. Meanwhile, the following considerations may be offered in its favour:

Logic states, and partly explains and applies, certain abstract principles which all other sciences take for granted; namely, the axioms above mentioned—the principles of Contradiction, of the Syllogism and of Causation. By exercising the student in the apprehension of these truths, and in the application of them to particular propositions, it educates the power of abstract thought. Every science is a model of method, a discipline in close and consecutive thinking; and this merit Logic ought to possess in a high degree.

For ages Logic has served as an introduction to Philosophy that is, to Metaphysics and speculative Ethics. It is of old and honourable descent: a man studies Logic in very good company. It is the warp upon which nearly the whole web of ancient, mediæval and modern Philosophy is woven. The history of thought is hardly intelligible without it.

As the science of proof, Logic gives an account of the *general* nature of evidence deductive and inductive, as applied in the physical and social sciences and in the affairs of life. The *general* nature of such evidence: it would be absurd of the logician to pretend to instruct the chemist, economist and merchant, as to the *special* character of the evidence requisite in their several spheres of judgment. Still, by investigating the general conditions of proof, he sets every man upon his guard against the insufficiency of evidence.

One application of the science of proof deserves special mention: namely, to that department of Rhetoric which has been the most developed, relating to persuasion by means of oratory, leader-writing, or pamphleteering. It is usually said that Logic is useful to convince the judgment, not to persuade the will: but one way of persuading the will is to convince the judgment that a certain course is advantageous; and although this is not always the readiest way, it is the most honourable, and leads to the most enduring results. Logic is the backbone of Rhetoric.

It has been disputed whether Logic is a science or an art; and, in fact, it may be considered in both ways. As a statement of general truths, of their relations to one another, and especially to the first principles, it is a science; but it is an art when, regarding truth as an end desired, it points out some of the means of attaining it—namely, to proceed by a regular method, to test every judgment by the principles of Logic, and to distrust whatever cannot be made consistent with them. Logic does not, in the first place, teach us to reason. We learn to reason as we learn to walk and talk, by the natural growth of our powers with some assistance from friends and neighbours. The way to develop one's power of reasoning is, first, to set oneself problems and try to solve them. Secondly, since the solving of a problem depends upon one's ability to call to mind parallel cases, one must learn as many facts as possible, and keep on learning all one's life; for nobody ever knew enough. Thirdly one must check all results by the principles of Logic. It is because of this checking, verifying, corrective function of Logic that it is sometimes called a Regulative or Normative Science. It cannot give any one originality or fertility of invention; but it enables us to check our inferences, revise our conclusions, and chasten the vagaries of ambitious speculation. It quickens our sense of bad reasoning both in others and in ourselves. A man who reasons deliberately, manages it better after studying Logic than he could before, if he is sincere about it and has common sense.

§ 5. The relation of Logic to other sciences:

(*a*) Logic is regarded by Spencer as co-ordinate with Mathematics, both being Abstract Sciences—that is, sciences of the *relations* in which things stand to one another, whatever the particular things may be that are so related; and this view seems to be, on the whole, just—subject, however, to qualifications that will appear presently.

Mathematics treats of the relations of all sorts of things considered as quantities, namely, as equal to, or greater or less than, one another. Things may be quantitatively equal or unequal in *degree*, as in comparing the temperature of bodies; or in *duration*; or in *spatial magnitude*, as with lines, superficies, solids; or in *number*. And it is assumed that the equality or inequality of things that cannot be directly compared, may be proved indirectly on the assumption that 'things equal to the same thing are equal,' etc.

Logic also treats of the relations of all sorts of things, but not as to their quantity. It considers (i) that one thing may be like or unlike another in certain attributes, as that iron is in many ways like tin or lead, and in many ways unlike carbon or sulphur: (ii) that attributes co-exist or coinhere (or do not) in the same subject, as metallic lustre, hardness, a certain atomic weight and a certain specific gravity coinhere in iron: and (iii) that one event follows another (or is the effect of it), as that the placing of iron in water causes it to rust. The relations of likeness and of coinherence are the ground of Classification; for it is by resemblance of coinhering attributes that things form classes: coinherence is the ground of judgments concerning Substance and Attribute, as that iron is metallic; and the relation of succession, in the mode of Causation, is the chief subject of the department of Induction. It is usual to group together these relations of attributes and of order in time, and call them qualitative, in order to contrast them with the quantitative relations which belong to Mathematics. And it is assumed that qualitative relations of things, when they cannot be directly perceived, may be proved indirectly by assuming the axiom of the Syllogism (chap. ix.) and the law of Causation (chap. xiv.).

So far, then, Logic and Mathematics appear to be co-ordinate and distinct sciences. But we shall see hereafter that the satisfactory treatment of that special order of events in time which constitutes Causation, requires a combination of Logic with Mathematics; and so does the treatment of Probability. And, again, Logic may be said to be, in a certain sense, 'prior to' or 'above' Mathematics as usually treated. For the Mathematics assume that one magnitude must be either equal or unequal to another, and that it cannot be both equal and unequal to it, and thus take for granted the principles of Contradiction and Excluded Middle; but the statement and elucidation of these Principles are left to Logic (chap. vi.). The Mathematics also classify and define magnitudes, as (in Geometry) triangles, squares, cubes, spheres; but the principles of classification and definition remain for Logic to discuss.

(*b*) As to the concrete Sciences, such as Astronomy, Chemistry, Zoology, Sociology—Logic (as well as Mathematics) is implied in them all; for all the propositions of which they consist involve causation, co-existence, and class-likeness. Logic is therefore said to be prior to them or above them: meaning by 'prior' not that it should be studied earlier, for that is not a good plan; meaning by 'above' not in dignity, for distinctions of dignity amongst liberal studies are absurd. But it is a philosophical idiom to call the abstract 'prior to,' or 'higher than,' the concrete (see Porphyry's Tree, chap. xxii. § 8); and Logic is more abstract than Astronomy or Sociology. Philosophy may thank that idiom for many a foolish notion.

(*c*) But, as we have seen, Logic does not investigate the truth, trustworthiness, or validity of its own principles; nor does Mathematics: this task belongs to Metaphysics, or Epistemology, the criticism of knowledge and beliefs.

Logic assumes, for example, that things are what to a careful scrutiny they seem to be; that animals, trees, mountains, planets, are bodies with various attributes, existing in space and changing in time; and that certain principles, such as Contradiction and Causation, are true of things and events. But Metaphysicians have raised many plausible objections to these assumptions. It has been urged that natural objects do not really exist on their own account, but only in dependence on some mind that contemplates them, and that even space and time are only our way of perceiving things; or, again, that although things do really exist on their own account, it is in an entirely different way from that in which we know them. As to the principle of Contradiction—that if an object has an attribute, it cannot at the same time and in the same way be without it (*e.g.*, if an animal is conscious, it is false that it is not conscious)—it has been contended that the speciousness of this principle is only due to the obtuseness of our minds, or even to the poverty of language, which cannot make the fine distinctions that exist in Nature. And as to Causation, it is sometimes doubted whether events always have physical causes; and it is often suggested that, granting they have physical causes, yet these are such as we can neither perceive nor conceive; belonging not to the order of Nature as we know it, but to the secret inwardness and reality of Nature, to the wells and reservoirs of power, not to the spray of the fountain that glitters in our eyes—'occult causes,' in short. Now these doubts and surmises are metaphysical spectres which it remains for Metaphysics to lay. Logic has no direct concern with them (although, of course, metaphysical discussion is expected to be logical), but keeps the plain path of plain beliefs, level with the comprehension of plain men. Metaphysics, as examining the grounds of Logic itself, is sometimes regarded as 'the higher Logic'; and, certainly, the study of Metaphysics is necessary to every one who would comprehend the nature and functions of Logic, or the place of his own mind and of Reason in the world.

(*d*) The relation of Logic to Psychology will be discussed in the next section.

(*e*) As a Regulative Science, pointing out the conditions of true inference (within its own sphere), Logic is co-ordinate with (i) Ethics, considered as assigning the conditions of right conduct, and with (ii) Æsthetics, considered as determining the principles of criticism and good taste.

§ 6. Three principal schools of Logicians are commonly recognised: Nominalist, Conceptualist, and Materialist, who differ as to what it is that Logic really treats of: the Nominalists say, 'of language'; the Conceptualists, 'of thought'; the Materialists, 'of relations of fact.' To illustrate these positions let us take authors who, if some of them are now neglected, have the merit of stating their contrasted views with a distinctness that later refinements tend to obscure.

(*a*) Whately, a well-known Nominalist, regarded Logic as the Science and Art of Reasoning, but at the same time as "entirely conversant about language"; that is to say, it is the business of Logic to discover those modes of statement which shall ensure the cogency of an argument, no matter what may be the subject under discussion. Thus, *All fish are cold-blooded*, ∴ *some cold-blooded things are fish:* this is a sound inference by the mere manner of expression; and equally sound is the inference, *All fish are warm-blooded*, ∴ *some warm-blooded things are fish*. The latter proposition may be false, but it follows; and (according to this doctrine) Logic is only concerned with the consistent use of words: the truth or falsity of the proposition itself is a question for Zoology. The short-coming of extreme Nominalism lies in speaking of language as if its meaning were unimportant. But Whately did not intend this: he was a man of great penetration and common-sense.

(*b*) Hamilton, our best-known Conceptualist, defined Logic as the science of the "formal laws of thought," and "of thought as thought," that is, without regard to the matter thought about. Just as Whately regarded Logic as concerned merely with cogent forms of statement, so Hamilton treated it as concerned merely with the necessary relations of thought. This doctrine is called Conceptualism, because the simplest element of thought is the Concept; that is, an abstract idea, such as is signified by the word *man, planet, colour, virtue*; not a representative or generic image, but the thought of all attributes common to any class of things. Men, planets, colours, virtuous actions or characters, have, severally, something in common on account of which they bear these general names; and the thought of what they have in common, as the ground of these names, is a Concept. To affirm or deny one concept of another, as *Some men are virtuous*, or *No man is perfectly virtuous*, is to form a Judgment, corresponding to the Proposition of which the other schools of Logic discourse. Conceptualism, then, investigates the conditions of consistent judgment.

To distinguish Logic from Psychology is most important in connection with Conceptualism. Concepts and Judgments being mental acts, or products of mental activity, it is often thought that Logic must be a department of Psychology. It is recognised of course, that Psychology deals with much more than Logic does, with sensation, pleasure and pain, emotion, volition; but in the region of the intellect, especially in its most deliberate and elaborate processes, namely, conception, judgment, and reasoning, Logic and Psychology seem to occupy common ground. In fact, however, the two sciences have little in common except a few general terms, and even these they employ in different senses. It is usual to point out that Psychology tries to explain the subjective *processes* of conception, judgment and reasoning, and to give their natural history; but that Logic is wholly concerned with the *results* of such processes, with concepts, judgments and reasonings, and merely with the validity of the results, that is, with their truth or consistency; whilst Psychology has nothing to do with their validity, but only with their causes. Besides, the logical judgment (in Formal Logic at least) is quite a different thing from the psychological: the latter involves feeling and belief, whereas the former is merely a given relation of concepts. S *is* P: that is a model logical judgment; there can be no question of believing it; but it is logically valid if M *is* P and S *is* M. When, again, in Logic, one deals with belief, it depends upon evidence; whereas, in Psychology belief is shown to depend upon causes which may have evidentiary value or may not; for Psychology explains quite impartially the growth of scientific insight and the growth of prejudice.

(*c*) Mill, Bain, and Venn are the chief Materialist logicians; and to guard against the error of confounding Materialism in Logic with the ontological doctrine that nothing exists but Matter, it may suffice to remember that in Metaphysics all these philosophers are Idealists. Materialism in Logic consists in regarding propositions as affirming or denying relations (*cf.* § 5) between matters-of-fact in the widest sense; not only physical facts, but ideas, social and moral relations; it consists, in short, in attending to the meaning of propositions. It treats the first principles of Contradiction and Causation as true of things so far as they are known to us, and not merely as conditions or tendencies of thought; and it takes these principles as conditions of right thinking, because they seem to hold good of Nature and human life.

To these differences of opinion it will be necessary to recur in the next chapter (§ 4); but here I may observe that it is easy to exaggerate their importance in Logic. There is really little at issue between schools of logicians as such, and as far as their doctrines run parallel; it is on the metaphysical grounds of their study, or as to its scope and comprehension, that they find a battle-field. The present work generally proceeds upon the third, or Materialist doctrine. If Deduction and Induction are regarded as mutually dependent parts of one science, uniting the discipline of consistent discourse with the method of investigating laws of physical phenomena, the Materialist doctrine, that the principles of Logic are founded on fact, seems to be the most natural way of thinking. But if the unity of Deduction and Induction is not disputed by the other schools, the Materialist may regard them as allies exhibiting in their own way the same body of truths. The Nominalist may certainly claim that his doctrine is indispensable: consistently cogent forms of statement are necessary both to the Conceptualist and to the Materialist; neither the relations of thought nor those of fact can be arrested or presented without the aid of language or some equivalent system of signs. The Conceptualist may urge that the Nominalist's forms of statement and argument exist for the sake of their meaning, namely, judgments and reasonings; and that the Materialist's laws of Nature are only judgments founded upon our conceptions of Nature; that the truth of observations and experiments depends upon our powers of perception; that perception is inseparable from understanding, and that a system of Induction may be constructed upon the axiom of Causation, regarded as a principle of Reason, just as well as by considering it as a law of Nature, and upon much the same lines. The Materialist, admitting all this, may say that a judgment is only the proximate meaning of a proposition, and that the ultimate meaning, the meaning of the judgment itself, is always some matter-of-fact; that the other schools have not hitherto been eager to recognise the unity of Deduction and Induction or to investigate the conditions of trustworthy experiments and observations within the limits of human understanding; that thought is itself a sort of fact, as complex in its structure, as profound in its relations, as subtle in its changes as any other fact, and therefore at least as hard to know; that to turn away from the full reality of thought in perception, and to confine Logic to artificially limited concepts, is to abandon the effort to push method to the utmost and to get as near truth as possible; and that as to Causation being a principle of Reason rather than of Nature, the distinction escapes his apprehension, since Nature seems to be that to which our private minds turn upon questions of Causation for correction and instruction; so that if he does not call Nature the Universal Reason, it is because he loves severity of style.

CHAPTER II
GENERAL ANALYSIS OF PROPOSITIONS

§ 1. Since Logic discusses the proof or disproof, or (briefly) the testing of propositions, we must begin by explaining their nature. A proposition, then, may first be described in the language of grammar as *a sentence indicative*; and it is usually expressed in the present tense.

It is true that other kinds of sentences, optative, imperative, interrogative, exclamatory, if they express or imply an assertion, are not beyond the view of Logic; but before treating such sentences, Logic, for greater precision, reduces them to their equivalent sentences indicative. Thus, *I wish it were summer* may be understood to mean, *The coming of summer is an object of my desire*. *Thou shalt not kill* may be interpreted as *Murderers are in danger of the judgment*. Interrogatories, when used in argument, if their form is affirmative, have negative force, and affirmative force if their form is negative. Thus, *Do hypocrites love virtue?* anticipates the answer, *No. Are not traitors the vilest of mankind?* anticipates the answer, *Yes*. So that the logical form of these sentences is, *Hypocrites are not lovers of virtue*; *Traitors are the vilest of mankind*. Impersonal propositions, such as *It rains*, are easily rendered into logical forms of equivalent meaning, thus: *Rain is falling*; or (if that be tautology), *The clouds are raining*. Exclamations may seem capricious, but are often part of the argument. *Shade of Chatham!* usually means *Chatham, being aware of our present foreign policy, is much disgusted*. It is in fact, an appeal to authority, without the inconvenience of stating what exactly it is that the authority declares.

§ 2. But even sentences indicative may not be expressed in the way most convenient to logicians. *Salt dissolves in water* is a plain enough statement; but the logician prefers to have it thus: *Salt is soluble in water*. For he says that a proposition is analysable into three elements: (1) a Subject (as *Salt*) about which something is asserted or denied; (2) a Predicate (as *soluble in water*) which is asserted or denied of the Subject, and (3) the Copula (*is* or *are*, or *is not* or *are not*), the sign of relation between the Subject and Predicate. The Subject and

Predicate are called the Terms of the proposition: and the Copula may be called the sign of predication, using the verb 'to predicate' indefinitely for either 'to affirm' or 'to deny.' Thus *S is P* means that the term *P* is given as related in some way to the term *S*. We may, therefore, further define a Proposition as 'a sentence in which one term is predicated of another.'

In such a proposition as *Salt dissolves*, the copula (*is*) is contained in the predicate, and, besides the subject, only one element is exhibited: it is therefore said to be *secundi adjacentis*. When all three parts are exhibited, as in *Salt is soluble*, the proposition is said to be *tertii adjacentis*.

For the ordinary purposes of Logic, in predicating attributes of a thing or class of things, the copula *is*, or *is not*, sufficiently represents the relation of subject and predicate; but when it is desirable to realise fully the nature of the relation involved, it may be better to use a more explicit form. Instead of saying *Salt—is—soluble*, we may say *Solubility—coinheres with—the nature of salt*, or *The putting of salt in water—is a cause of—its dissolving*: thus expanding the copula into a full expression of the relation we have in view, whether coinherence or causation.

§ 3. The sentences of ordinary discourse are, indeed, for the most part, longer and more complicated than the logical form of propositions; it is in order to prove them, or to use them in the proof of other propositions, that they are in Logic reduced as nearly as possible to such simple but explicit expressions as the above (*tertii adjacentis*). A Compound Proposition, reducible to two or more simple ones, is said to be exponible.

The modes of compounding sentences are explained in every grammar-book. One of the commonest forms is the copulative, such as *Salt is both savoury and wholesome*, equivalent to two simple propositions: *Salt is savoury; Salt is wholesome. Pure water is neither sapid nor odorous*, equivalent to *Water is not sapid; Water is not odorous*. Or, again, *Tobacco is injurious, but not when used in moderation*, equivalent to *Much tobacco is injurious; a little is not*.

Another form of Exponible is the Exceptive, as *Kladderadatsch is published daily, except on week-days*, equivalent to *Kladderadatsch is published on Sunday; it is not published any other day*. Still another Exponible is the Exclusive, as *Only men use fire*, equivalent to *Men are users of fire; No other animals are*. Exceptive and exclusive sentences are, however, equivalent forms; for we may say, *Kladderadatsch is published only on Sunday*; and *No animals use fire, except men*.

There are other compound sentences that are not exponible, since, though they contain two or more verbal clauses, the construction shows that these are inseparable. Thus, *If cats are scarce, mice are plentiful*, contains two verbal clauses; but *if cats are scarce* is conditional, not indicative; and *mice are plentiful* is subject to the condition that *cats are scarce*. Hence the whole sentence is called a Conditional Proposition. For the various forms of Conditional Propositions see chap. v. § 4.

But, in fact, to find the logical force of recognised grammatical forms is the least of a logician's difficulties in bringing the discourses of men to a plain issue. Metaphors, epigrams, innuendoes and other figures of speech present far greater obstacles to a lucid reduction whether for approval or refutation. No rules can be given for finding everybody's meaning. The poets have their own way of expressing themselves; sophists, too, have their own way. And the point often lies in what is unexpressed. Thus, "barbarous nations make, the civilised write history," means that civilised nations do not make history, which none is so brazen as openly to assert. Or, again, "Alcibiades is dead, but X is still with us"; the whole meaning of this 'exponible' is that X would be the lesser loss to society. Even an epithet or a suffix may imply a proposition: *This personage* may mean *X is a pretentious nobody*.

How shall we interpret such illusive predications except by cultivating our literary perceptions, by reading the most significant authors until we are at home with them? But, no doubt, to disentangle the compound propositions, and to expand the abbreviations of literature and conversation, is a useful logical exercise. And if it seem a laborious task thus to reduce to its logical elements a long argument in a speech or treatise, it should be observed that, as a rule, in a long discourse only a few sentences are of principal importance to the reasoning, the rest being explanatory or illustrative digression, and that a close scrutiny of these cardinal sentences will frequently dispense us from giving much attention to the rest.

§ 4. But now, returning to the definition of a Proposition given in § 2, that it is 'a sentence in which one term is predicated of another,' we must consider what is the import of such predication. For the definition, as it stands, seems to be purely Nominalist. Is a proposition nothing more than a certain synthesis of words; or, is it meant to correspond with something further, a synthesis of ideas, or a relation of facts?

Conceptualist logicians, who speak of judgments instead of propositions, of course define the judgment in their own language. According to Hamilton, it is "a recognition of the relation of congruence or conflict in which two concepts stand to each other." To lighten the sentence, I have omitted one or two qualifications (Hamilton's *Lectures on Logic*, xiii.). "Thus," he goes on "if we compare the thoughts *water*, *iron*, and *rusting*, we find them congruent, and connect them into a single thought, thus: *water rusts iron*—in that case we form a judgment." When a judgment is expressed in words, he says, it is called a proposition.

But has a proposition no meaning beyond the judgment it expresses? Mill, who defines it as "a portion of discourse in which a predicate is affirmed or denied of a subject" (*Logic*, Book 1., chap. iv. § 1.), proceeds to inquire into the import of propositions (Book 1., chap. v.), and finds three classes of them: (*a*) those in which one proper name is predicated of another; and of these Hobbes's Nominalist definition is adequate, namely, that a proposition asserts or denies that the predicate is a name for the same thing as the subject, as *Tully is Cicero*.

(*b*) Propositions in which the predicate means a part (or the whole) of what the subject means, as *Horses are animals*, *Man is a rational animal*. These are Verbal Propositions (see below: chap. v. § 6), and their import consists in affirming or denying a coincidence between the meanings of names, as *The meaning of 'animal' is part of the meaning of 'horse.'* They are partial or complete definitions.

But (*c*) there are also Real Propositions, whose predicates do not mean the same as their subjects, and whose import consists in affirming or denying one of five different kinds of matter of fact: (1) That the subject exists, or does not; as if we say *The bison exists*, *The great auk is extinct*. (2) Co-existence, as *Man is mortal*; that is, *the being subject to death coinheres with the qualities on account of which we call certain objects men*. (3) Succession, as *Night follows day*. (4) Causation (a particular kind of Succession), as *Water rusts iron*. (5) Resemblance, as *The colour of this geranium is like that of a soldier's coat*, or $A = B$.

On comparing this list of real predications with the list of logical relations given above (chap. i. § 5 (*a*)), it will be seen that the two differ only in this, that I have there omitted simple Existence. Nothing simply exists, unrelated either in Nature or in knowledge. Such a proposition as *The bison exists* may, no doubt, be used in Logic (subject to interpretation) for the sake of custom or for the sake of brevity; but it means that some specimens are still to be found in N. America, or in Zoological gardens.

Controversy as to the Import of Propositions really turns upon a difference of opinion as to the scope of Logic and the foundations of knowledge. Mill was dissatisfied with the "congruity" of concepts as the basis of a judgment. Clearly, mere congruity does not justify belief. In the proposition *Water rusts iron*, the concepts *water, rust* and *iron* may be congruous, but does any one assert their connection on that ground? In the proposition *Murderers are haunted by the ghosts of their victims*, the concepts *victim, murderer, ghost* have a high degree of congruity; yet, unfortunately, I cannot believe it: there seems to be no such cheap defence of innocence. Now, Mill held that Logic is concerned with the grounds of belief, and that the scope of Logic includes Induction as well as Deduction; whereas, according to Hamilton, Induction is only Modified Logic, a mere appendix to the theory of the "forms of thought as thought." Indeed, Mill endeavoured in his *Logic* to probe the grounds of belief deeper than usual, and introduced a good deal of Metaphysics—either too much or not enough—concerning the ground of axioms. But, at any rate, his great point was that belief, and therefore (for the most part) the Real Proposition, is concerned not merely with the relations of words, or even of ideas, but with matters of fact; that is, both propositions and judgments point to something further, to the relations of things which we can examine, not merely by thinking about them (comparing them in thought), but by observing them with the united powers of thought and perception. This is what convinces us that *water rusts iron*: and the difficulty of doing this is what prevents our feeling sure that *murderers are haunted by the ghosts of their victims*. Hence, although Mill's definition of a proposition, given above, is adequate for propositions in general; yet that kind of proposition (the Real) with regard to which Logic (in Mill's view) investigates the conditions of proof, may be more explicitly and pertinently defined as 'a predication concerning the relation of matters of fact.'

§ 5. This leads to a very important distinction to which we shall often have to refer in subsequent pages—namely, the distinction between the Form and the Matter of a proposition or of an argument. The distinction between Form and Matter, as it is ordinarily employed, is easily understood. An apple growing in the orchard and a waxen apple on the table may have the same shape or form, but they consist of different materials; two real apples may have the same shape, but contain distinct ounces of apple-stuff, so that after one is eaten the other remains to be eaten. Similarly, tables may have the same shape, though one be made of marble, another of oak, another of iron. The form is common to several things, the matter is peculiar to each. Metaphysicians have carried the distinction further: apples, they say, may have not only the same outward shape, but the same inward constitution, which, therefore, may be called the Form of apple-stuff itself—namely, a certain pulpiness, juiciness, sweetness, *etc.*; qualities common to all dessert apples: yet their Matter is different, one being here, another there—differing in place or time, if in nothing else. The definition of a species is the form of every specimen of it.

To apply this distinction to the things of Logic: it is easy to see how two propositions may have the same Form but different Matter: not using 'Form' in the sense of 'shape,' but for that which is common to many things, in contrast with that which is peculiar to each. Thus, *All male lions are tawny* and *All water is liquid at 50° Fahrenheit*, are two propositions that have the same form, though their matter is entirely different. They both predicate something of the whole of their subjects, though their subjects are different, and so are the things predicated of them. Again, *All male lions have tufted tails* and *All male lions have manes*, are two propositions having the same form and, in their subjects, the same matter, but different matter in their predicates. If, however, we take two such propositions as these: *All male lions have manes* and *Some male lions have manes*, here the matter is the same in both, but the form is different—in the first, predication is made concerning *every* male lion; in the second of only *some* male lions; the first is *universal*, the second is *particular*. Or, again, if we take *Some tigers are man-eaters* and *Some tigers are not man-eaters*, here too the matter is the same, but the form is different; for the first proposition is *affirmative*, whilst the second is *negative*.

§ 6. Now, according to Hamilton and Whately, pure Logic has to do only with the Form of propositions and arguments. As to their Matter, whether they are really true in fact, that is a question, they said, not for Logic, but for experience, or for the special sciences. But Mill desired so to extend logical method as to test the material truth of propositions: he thought that he could expound a method by which experience itself and the conclusions of the special sciences may be examined.

To this method it may be objected, that the claim to determine Material Truth takes for granted that the order of Nature will remain unchanged, that (for example) water not only at present is a liquid at 50° Fahrenheit, but will always be so; whereas (although we have no reason to expect such a thing) the order of Nature may alter—it is at least supposable—and in that event water may freeze at such a temperature. Any matter of fact, again, must depend on observation, either directly, or by inference—as when something is asserted about atoms or ether. But observation and material inference are subject to the limitations of our faculties; and however we may aid observation by microscopes and micrometers, it is still observation; and however we may correct our observations by repetition, comparison and refined mathematical methods of making allowances, the correction of error is only an approximation to accuracy. Outside of Formal Reasoning, suspense of judgment is your only attitude.

But such objections imply that nothing short of absolute truth has any value; that all our discussions and investigations in science or social affairs are without logical criteria; that Logic must be confined to symbols, and considered entirely as mental gymnastics. In this book prominence will be given to the character of Logic as a formal science, and it will also be shown that Induction itself may be treated formally; but it will be assumed that logical forms are valuable as representing the actual relations of natural and social phenomena.

§ 7. Symbols are often used in Logic instead of concrete terms, not only in Symbolic Logic where the science is treated algebraically (as by Dr. Venn in his *Symbolic Logic*), but in ordinary manuals; so that it may be well to explain the use of them before going further.

It is a common and convenient practice to illustrate logical doctrines by examples: to show what is meant by a Proposition we may give *salt is soluble*, or *water rusts iron*: the copulative exponible is exemplified by *salt is savoury and wholesome*; and so on. But this procedure has some disadvantages: it is often cumbrous; and it may distract the reader's attention from the point to be explained by exciting his interest in the special fact of the illustration. Clearly, too, so far as Logic is formal, no particular matter of fact can adequately illustrate any of its doctrines. Accordingly, writers on Logic employ letters of the alphabet instead of concrete terms, (say) *X* instead of *salt* or instead of *iron*, and (say) *Y* instead of *soluble* or instead of *rusted by water*; and then a proposition may be represented by *X is Y*. It is still more usual to represent a proposition by *S is (or is not) P*, *S* being the initial of Subject and *P* of Predicate; though this has the drawback that if we argue—*S is P*, therefore *P is S*, the symbols in the latter proposition no longer have the same significance, since the former subject is now the predicate.

Again, negative terms frequently occur in Logic, such as *not-water*, or *not-iron*, and then if *water* or *iron* be expressed by *X*, the corresponding negative may be expressed by *x*; or, generally, if a capital letter stand for a positive term, the corresponding small letter represents the negative. The same device may be adopted to express contradictory terms: either of them being *X*, the other is *x* (see chap. iv., §§ 7-8); or the contradictory terms may be expressed by x and $\bar{x}$, y and $\bar{y}$.

And as terms are often compounded, it may be convenient to express them by a combination of letters: instead of illustrating such a case by *boiling water* or *water that is boiling*, we may write *XY*; or since positive and negative terms may be compounded, instead of illustrating this by *water that is not boiling*, we may write *Xy*.

The convenience of this is obvious; but it is more than convenient; for, if one of the uses of Logic be to discipline the power of abstract thought, this can be done far more effectually by symbolic than by concrete examples; and if such discipline were the only use of Logic it might be best to discard concrete illustrations altogether, at least in advanced text-books, though no doubt the practice would be too severe for elementary manuals. On the other hand, to show the practical applicability of Logic to the arguments and proofs of actual life, or even of the concrete sciences, merely symbolic illustration may be not only useless but even misleading. When we speak of politics, or poetry, or species, or the weather, the terms that must be used can rarely have the distinctness and isolation of X and Y; so that the perfunctory use of symbolic illustration makes argument and proof appear to be much simpler and easier matters than they really are. Our belief in any proposition never rests on the proposition itself, nor merely upon one or two others, but upon the immense background of our general knowledge and beliefs, full of circumstances and analogies, in relation to which alone any given proposition is intelligible. Indeed, for this reason, it is impossible to illustrate Logic sufficiently: the reader who is in earnest about the cogency of arguments and the limitation of proofs, and is scrupulous as to the degrees of assent that they require, must constantly look for illustrations in his own knowledge and experience and rely at last upon his own sagacity.

CHAPTER III
OF TERMS AND THEIR DENOTATION

§ 1. In treating of Deductive Logic it is usual to recognise three divisions of the subject: first, the doctrine of Terms, words, or other signs used as subjects or predicates; secondly, the doctrine of Propositions, analysed into terms related; and, thirdly, the doctrine of the Syllogism in which propositions appear as the grounds of a conclusion.

The terms employed are either letters of the alphabet, or the words of common language, or the technicalities of science; and since the words of common language are most in use, it is necessary to give some account of common language as subserving the purposes of Logic. It has been urged that we cannot think or reason at all without words, or some substitute for them, such as the signs of algebra; but this is an exaggeration. Minds greatly differ, and some think by the aid of definite and comprehensive picturings, especially in dealing with problems concerning objects in space, as in playing chess blindfold, inventing a machine, planning a tour on an imagined map. Most people draw many simple inferences by means of perceptions, or of mental imagery. On the other hand, some men think a good deal without any continuum of words and without any imagery, or with none that seems relevant to the purpose. Still the more elaborate sort of thinking, the grouping and concatenation of inferences, which we call reasoning, cannot be carried far without language or some equivalent system of signs. It is not merely that we need language to express our reasonings and communicate them to others: in solitary thought we often depend on words—'talk to ourselves,' in fact; though the words or sentences that then pass through our minds are not always fully formed or articulated. In Logic, moreover, we have carefully to examine the grounds (at least the proximate grounds) of our conclusions; and plainly this cannot be done unless the conclusions in question are explicitly stated and recorded.

Conceptualists say that Logic deals not with the process of thinking (which belongs to Psychology) but with its results; not with conceiving but with concepts; not with judging but with judgments. Is the concept self-consistent or adequate? Logic asks; is the judgment capable of proof? Now, it is only by recording our thoughts in language that it becomes possible to distinguish between the process and the result of thought. Without language, the act and the product of thinking would be identical and equally evanescent. But by carrying on the process in language and remembering or otherwise recording it, we obtain a result which may be examined according to the principles of Logic.

§ 2. As Logic, then, must give some account of language, it seems desirable to explain how its treatment of language differs from that of Grammar and from that of Rhetoric.

Grammar is the study of the words of some language, their classification and derivation, and of the rules of combining them, according to the usage at any time recognised and followed by those who are considered correct writers or speakers. Composition may be faultless in its grammar, though dull and absurd.

Rhetoric is the study of language with a view to obtaining some special effect in the communication of ideas or feelings, such as picturesqueness in description, vivacity in narration, lucidity in exposition, vehemence in persuasion, or literary charm. Some of these ends are often gained in spite of faulty syntax or faulty logic; but since the few whom bad grammar saddens or incoherent arguments divert are not carried away, as they else might be, by an unsophisticated orator, Grammar and Logic are necessary to the perfection of Rhetoric. Not that Rhetoric is in bondage to those other sciences; for foreign idioms and such figures as the ellipsis, the anacoluthon, the oxymoron, the hyperbole, and violent inversions have their place in the magnificent style; but authors unacquainted with Grammar and Logic are not likely to place such figures well and wisely. Indeed, common idioms, though both grammatically and rhetorically justifiable, both correct and effective, often seem illogical. 'To fall asleep,' for example, is a perfect English phrase; yet if we examine severally the words it consists of, it may seem strange that their combination should mean anything at all.

But Logic only studies language so far as necessary in order to state, understand, and check the evidence and reasonings that are usually embodied in language. And as long as meanings are clear, good Logic is compatible with false concords and inelegance of style.

§ 3. Terms are either Simple or Composite: that is to say, they may consist either of a single word, as 'Chaucer,' 'civilisation'; or of more than one, as 'the father of English poetry,' or 'modern civilised nations.' Logicians classify words according to their uses in forming propositions; or, rather, they classify the uses of words as terms, not the words themselves; for the same word may fall into different classes of terms according to the way in which it is used. (*Cf.* Mr. Alfred Sidgwick's *Distinction and the Criticism of Beliefs*, chap. xiv.)

Thus words are classified as Categorematic or Syncategorematic. A word is Categorematic if used singly as a term without the support of other words: it is Syncategorematic when joined with other words in order to constitute the subject or predicate of a proposition. If we say *Venus is a planet whose orbit is inside the Earth's*, the subject, 'Venus,' is a word used categorematically as a simple term; the predicate is a composite term whose constituent words (whether substantive, relative, verb, or preposition) are used syncategorematically.

Prepositions, conjunctions, articles, adverbs, relative pronouns, in their ordinary use, can only enter into terms along with other words having a substantive, adjectival or participial force; but when they are themselves the things spoken of and are used substantively (*suppositio materialis*), they are categorematic. In the proposition, *'Of' was used more indefinitely three hundred years ago than it is now*, 'of' is categorematic. On the other hand, all substantives may be used categorematically; and the same self-sufficiency is usually recognised in adjectives and participles. Some, however, hold that the categorematic use of adjectives and participles is due to an ellipsis which the logician should fill up; that instead of *Gold is heavy*, he should say *Gold is a heavy metal*; instead of *The sun is shining*, *The sun is a body shining*. But in these cases the words 'metal' and 'body' are unmistakable tautology, since 'metal' is implied in gold and 'body' in sun. But, as we have seen, any of these kinds of word, substantive, adjective, or participle, may occur syncategorematically in connection with others to form a composite term.

§ 4. Most terms (the exceptions and doubtful cases will be discussed hereafter) have two functions, a denotative and a connotative. A term's denotative function is, to be the name or sign of something or some multitude of things, which are said to be called or denoted by the term. Its connotative function is, to suggest certain qualities and characteristics of the things denoted, so that it cannot be used literally as the name of any other things; which qualities and characteristics are said to be implied or connoted by the term. Thus 'sheep' is the name of certain animals, and its connotation prevents its being used of any others. That which a term directly indicates, then, is its *Denotation*; that sense or customary use of it which limits the Denotation is its *Connotation* (ch. iv.). Hamilton and others use 'Extension' in the sense of Denotation, and 'Intension' or 'Comprehension' in the sense of Connotation. Now, terms may be classified, first according to what they stand for or denote; that is, according to their *Denotation*. In this respect, the use of a term is said to be either Concrete or Abstract.

A term is Concrete when it denotes a 'thing'; that is, any person, object, fact, event, feeling or imagination, considered as capable of having (or consisting of) qualities and a determinate existence. Thus 'cricket ball' denotes any object having a certain size, weight, shape, colour, *etc.* (which are its qualities), and being at any given time in some place and related to other objects—in the bowler's hands, on the grass, in a shop window. Any 'feeling of heat' has a certain intensity, is pleasurable or painful, occurs at a certain time, and affects some part or the whole of some animal. An imagination, indeed (say, of a fairy), cannot be said in the same sense to have locality; but it depends on the thinking of some man who has locality, and is definitely related to his other thoughts and feelings.

A term is Abstract, on the other hand, when it denotes a quality (or qualities), considered by itself and without determinate existence in time, place, or relation to other things. 'Size,' 'shape,' 'weight,' 'colour,' 'intensity,' 'pleasurableness,' are terms used to denote such qualities, and are then abstract in their denotation. 'Weight' is not something with a determinate existence at a given time; it exists not merely in some particular place, but wherever there is a heavy thing; and, as to relation, at the same moment it combines in iron with solidity and in mercury with liquidity. In fact, a quality is a point of agreement in a multitude of different things; all heavy things agree in weight, all round things in roundness, all red things in redness; and an abstract term denotes such a point (or points) of agreement among the things denoted by concrete terms. Abstract terms result from the analysis of concrete things into their qualities; and conversely a concrete term may be viewed as denoting the synthesis of qualities into an individual thing. When several things agree in more than one quality, there may be an abstract term denoting the union of qualities in which they agree, and omitting their peculiarities; as 'human nature' denotes the common qualities of men, 'civilisation' the common conditions of civilised peoples.

Every general name, if used as a concrete term, has, or may have, a corresponding abstract term. Sometimes the concrete term is modified to form the abstract, as 'greedy—greediness'; sometimes a word is adapted from another language, as 'man—humanity'; sometimes a composite term is used, as 'mercury—the nature of mercury,' *etc.* The same concrete may have several abstract correlatives, as 'man—manhood, humanity, human nature'; 'heavy—weight, gravity, ponderosity'; but in such cases the abstract terms are not used quite synonymously; that is, they imply different ways of considering the concrete.

Whether a word is used as a concrete or abstract term is in most instances plain from the word itself, the use of most words being pretty regular one way or the other; but sometimes we must judge by the context. 'Weight' may be used in the abstract for 'gravity,' or in the concrete for a measure; but in the latter sense it is syncategorematic (in the singular), needing at least the article 'a (or the) weight.' 'Government' may mean 'supreme political authority,' and is then abstract; or, the men who happen to be ministers, and is then concrete; but in this case, too, the article is usually prefixed. 'The life' of any man may mean his vitality (abstract), as in "Thus following life in creatures we dissect"; or, the series of events through which he passes (concrete), as in 'the life of Nelson as narrated by Southey.'

It has been made a question whether the denotation of an abstract term may itself be the subject of qualities. Apparently 'weight' may be greater or less, 'government' good or bad, 'vitality' intense or dull. But if every subject is modified by a quality, a quality is also modified by making it the subject of another; and, if so, it seems then to become a new quality. The compound terms 'great weight,' 'bad government,' 'dull vitality,' have not the same denotation as the simple terms 'weight, 'government,' 'vitality': they imply, and may be said to connote, more special concrete experience, such as the effort felt in lifting a trunk, disgust at the conduct of officials, sluggish movements of an animal when irritated. It is to such concrete experiences that we have always to refer in order fully to realise the meaning of abstract terms, and therefore, of course, to understand any qualification of them.

§ 5. Concrete terms may be subdivided according to the number of things they denote and the way in which they denote them. A term may denote one thing or many: if one, it is called Singular; if many, it may do so distributively, and then it is General; or, as taken all together, and then it is Collective: one, then; any one of many; many in one.

Among Singular Terms, each denoting a single thing, the most obvious are Proper Names, such as Gibraltar or George Washington, which are merely marks of individual things or persons, and may form no part of the common language of a country. They are thus distinguished from other Singular Terms, which consist of common words so combined as to restrict their denotation to some individual, such as, 'the strongest man on earth.'

Proper Terms are often said to be arbitrary signs, because their use does not depend upon any reason that may be given for them. Gibraltar had a meaning among the Moors when originally conferred; but no one now knows what it was, unless he happens to have learned it; yet the name serves its purpose as well as if it were "Rooke's Nest." Every Newton or Newport year by year grows old, but to alter the name would cause only confusion. If such names were given by mere caprice it would make no difference; and they could not be more cumbrous, ugly, or absurd than many of those that are given 'for reasons.'

The remaining kinds of Singular Terms are drawn from the common resources of the language. Thus the pronouns 'he,' 'she,' 'it,' are singular terms, whose present denotation is determined by the occasion and context of discourse: so with demonstrative phrases—'the man,' 'that horse.' Descriptive names may be more complex, as 'the wisest man of Gotham,' which is limited to some individual by the superlative suffix; or 'the German Emperor,' which is limited by the definite article—the general term 'German Emperor' being thereby restricted either to the reigning monarch or to the one we happen to be discussing. Instead of the definite, the indefinite article may be used to make general terms singular, as 'a German Emperor was crowned at Versailles' (*individua vaga*).

Abstract Terms are ostensively singular: 'whiteness' (*e.g.*) is one quality. But their full meaning is general: 'whiteness' stands for all white things, so far as white. Abstract terms, in fact, are only formally singular.

General Terms are words, or combinations of words, used to denote any one of many things that resemble one another in certain respects. 'George III.' is a Singular Term denoting one man; but 'King' is a General Term denoting him and all other men of the same rank; whilst the compound 'crowned head' is still more general, denoting kings and also emperors. It is the nature of a general term, then, to be used in the same sense of whatever it denotes; and its most characteristic form is the Class-name, whether of objects, such as 'king,' 'sheep,' 'ghost'; or of events, such as 'accession,' 'purchase,' 'manifestation.' Things and events are known by their qualities and relations; and every such aspect, being a point of resemblance to some other things, becomes a ground of generalisation, and therefore a ground for the need and use of general terms. Hence general terms are far the most important sort of terms in Logic, since in them general propositions are expressed and, moreover (with rare exceptions), all predicates are general. For, besides these typical class-names, attributive words are general terms, such as 'royal,' 'ruling,' 'woolly,' 'bleating,' 'impalpable,' 'vanishing.'

Infinitives may also be used as general terms, as '*To err is human*'; but for logical purposes they may have to be translated into equivalent substantive forms, as *Foolish actions are characteristic of mankind.* Abstract terms, too, are (as I observed) equivalent to general terms; 'folly' is abstract for 'foolish actions.' '*Honesty is the best policy*' means *people who are honest may hope to find their account in being so;* that is, in the effects of their honest actions, provided they are wise in other ways, and no misfortunes attend them. The abstract form is often much the more succinct and forcible, but for logical treatment it needs to be interpreted in the general form.

By antonomasia proper names may become general terms, as if we say '*A Johnson' would not have written such a book—i.e.*, any man of his genius for elaborate eloquence.

A Collective Term denotes a multitude of similar things considered as forming one whole, as 'regiment,' 'flock,' 'nation': not distributively, that is, not the similar things severally; to denote them we must say 'soldiers of the regiment,' 'sheep of the flock,' and so on. If in a multitude of things there is no resemblance, except the fact of being considered as parts of one whole, as 'the world,' or 'the town of Nottingham' (meaning its streets and houses, open spaces, people, and civic organisation), the term denoting them as a whole is Singular; but 'the world' or 'town of Nottingham,' meaning the inhabitants only, is Collective.

In their strictly collective use, all such expressions are equivalent to singular terms; but many of them may also be used as general terms, as when we speak of 'so many regiments of the line,' or discuss the 'plurality of worlds'; and in this general use they denote any of a multitude of things of the same kind—regiments, or habitable worlds.

Names of substances, such as 'gold,' 'air,' 'water,' may be employed as singular, collective, or general terms; though, perhaps, as singular terms only figuratively, as when we say *Gold is king*. If we say with Thales, '*Water is the source of all things*,' 'water' seems to be used collectively. But substantive names are frequently used as general terms. For example, *Gold is heavy* means 'in comparison with other things,' such as water. And, plainly, it does not mean that the aggregate of gold is heavier than the aggregate of water, but only that its specific gravity is greater; that is, bulk for bulk, any piece of gold is heavier than water.

Finally, any class-name may be used collectively if we wish to assert something of the things denoted by it, not distributively but altogether, as that *Sheep are more numerous than wolves.*

CHAPTER IV
THE CONNOTATION OF TERMS

§ 1. Terms are next to be classified according to their Connotation—that is, according to what they imply as characteristic of the things denoted. We have seen that general names are used to denote many things in the same sense, because the things denoted resemble one another in certain ways: it is this resemblance in certain points that leads us to class the things together and call them by the same name; and therefore the points of resemblance constitute the sense or meaning of the name, or its Connotation, and limit its applicability to such things as have these characteristic qualities. 'Sheep' for example, is used in the same sense, to denote any of a multitude of animals that resemble one another: their size, shape, woolly coats, cloven hoofs, innocent ways and edibility are well known. When we apply to anything the term 'sheep,' we imply that it has these qualities: 'sheep,' denoting the animal, connotes its possessing these characteristics; and, of course, it cannot, without a figure of speech or a blunder, be used to denote anything that does not possess all these qualities. It is by a figure of speech that the term 'sheep' is applied to some men; and to apply it to goats would be a blunder.

Most people are very imperfectly aware of the connotation of the words they use, and are guided in using them merely by the custom of the language. A man who employs a word quite correctly may be sadly posed by a request to explain or define it. Moreover, so far as we are aware of the connotation of terms, the number and the kind of attributes we think of, in any given case, vary with the depth of our interest, and with the nature of our interest in the things denoted. 'Sheep' has one meaning to a touring townsman, a much fuller one to a farmer, and yet a different one to a zoologist. But this does not prevent them agreeing in the use of the word, as long as the qualities they severally include in its meaning are not incompatible.

All general names, and therefore not only class-names, like 'sheep,' but all attributives, have some connotation. 'Woolly' denotes anything that bears wool, and connotes the fact of bearing wool; 'innocent' denotes anything that habitually and by its disposition does no harm (or has not been guilty of a particular offence), and connotes a harmless character (or freedom from particular guilt); 'edible' denotes whatever can be eaten with good results, and connotes its suitability for mastication, deglutition, digestion, and assimilation.

§ 2. But whether all terms must connote as well as denote something, has been much debated. Proper names, according to what seems the better opinion, are, in their ordinary use, not connotative. To say that they have no meaning may seem violent: if any one is called John Doe, this name, no doubt, means a great deal to his friends and neighbours, reminding them of his stature and physiognomy, his air and gait, his wit and wisdom, some queer stories, and an indefinite number of other things. But all this significance is local or accidental; it only exists for those who know the individual or have heard him described: whereas a general name gives information about any thing or person it denotes to everybody who understands the language, without any particular knowledge of the individual.

We must distinguish, in fact, between the peculiar associations of the proper name and the commonly recognised meaning of the general name. This is why proper names are not in the dictionary. Such a name as London, to be sure, or Napoleon Buonaparte, has a significance not merely local; still, it is accidental. These names are borne by other places and persons than those that have rendered them famous. There are Londons in various latitudes, and, no doubt, many Napoleon Buonapartes in Louisiana; and each name has in its several denotations an altogether different suggestiveness. For its suggestiveness is in each application determined by the peculiarities of the place or person denoted; it is not given to the different places (or to the different persons) because they have certain characteristics in common.

However, the scientific grounds of the doctrine that proper names are non-connotative, are these: The peculiarities that distinguish an individual person or thing are admitted to be infinite, and anything less than a complete enumeration of these peculiarities may fail to distinguish and identify the individual. For, short of a complete enumeration of them, the description may be satisfied by two or more individuals; and in that case the term denoting them, if limited by such a description, is not a proper but a general name, since it is applicable to two or more in the same sense. The existence of other individuals to whom it applies may be highly improbable; but, if it be logically possible, that is enough. On the other hand, the enumeration of infinite peculiarities is certainly impossible. Therefore proper names have no assignable connotation. The only escape from this reasoning lies in falling back upon time and place, the principles of individuation, as constituting the connotation of proper names. Two things cannot be at the same time in the same place: hence 'the man who was at a certain spot on the bridge of Lodi at a certain instant in a certain year' suffices to identify Napoleon Buonaparte for that instant. Supposing no one else to have borne the name, then, is this its connotation? No one has ever thought so. And, at any rate, time and place are only extrinsic determinations (suitable indeed to events like the battle of Lodi, or to places themselves like London); whereas the connotation of a general term, such as 'sheep,' consists of intrinsic qualities. Hence, then, the scholastic doctrine 'that individuals have no essence' (see chap. xxii. § 9), and Hamilton's dictum 'that every concept is inadequate to the individual,' are justified.

General names, when used as proper names, lose their connotation, as Euxine or Newfoundland.

Singular terms, other than Proper, have connotation; either in themselves, like the singular pronouns 'he,' 'she,' 'it,' which are general in their applicability, though singular in application; or, derivatively, from the general names that combine to form them, as in 'the first Emperor of the French' or the 'Capital of the British Empire.'

§ 3. Whether Abstract Terms have any connotation is another disputed question. We have seen that they denote a quality or qualities of something, and that is precisely what general terms connote: 'honesty' denotes a quality of some men; 'honest' connotes the same quality, whilst denoting the men who have it.

The denotation of abstract terms thus seems to exhaust their force or meaning. It has been proposed, however, to regard them as connoting the qualities they directly stand for, and not denoting anything; but surely this is too violent. To denote something is the same as to be the name of something (whether real or unreal), which every term must be. It is a better proposal to regard their denotation and connotation as coinciding; though open to the objection that 'connote' means 'to mark along with' something else, and this plan leaves nothing else. Mill thought that abstract terms are connotative when, besides denoting a quality, they suggest a quality of that quality (as 'fault' implies 'hurtfulness'); but against this it may be urged that one quality cannot bear another, since every qualification of a quality constitutes a distinct quality in the total ('milk-whiteness' is distinct from 'whiteness,' cf. chap. iii. § 4). After all, if it is the most consistent plan, why not say that abstract, like proper, terms have no connotation?

But if abstract terms must be made to connote something, should it not be those things, indefinitely suggested, to which the qualities belong? Thus 'whiteness' may be considered to connote either snow or vapour, or any white thing, apart from one or other of which the quality has no existence; whose existence therefore it implies. By this course the denotation and connotation of abstract and of general names would be exactly reversed. Whilst the denotation of a general name is limited by the qualities connoted, the connotation of an abstract name includes all the things in which its denotation is realised. But the whole difficulty may be avoided by making it a rule to translate, for logical purposes, all abstract into the corresponding general terms.

§ 4. If we ask how the connotation of a term is to be known, the answer depends upon how it is used. If used scientifically, its connotation is determined by, and is the same as, its definition; and the definition is determined by examining the things to be denoted, as we shall see in chap. xxii. If the same word is used as a term in different sciences, as 'property' in Law and in Logic, it will be differently defined by them, and will have, in each use, a correspondingly different connotation. But terms used in popular discourse should, as far as possible, have their connotations determined by classical usage, *i.e.*, by the sense in which they are used by writers and speakers who are acknowledged masters of the language, such as Dryden and Burke. In this case the classical connotation determines the definition; so that to define terms thus used is nothing else than to analyse their accepted meanings.

It must not, however, be supposed that in popular use the connotation of any word is invariable. Logicians have attempted to classify terms into Univocal (having only one meaning) and Æquivocal (or ambiguous); and no doubt some words (like 'civil,' 'natural,' 'proud,' 'liberal,' 'humorous') are more manifestly liable to ambiguous use than some others. But in truth all general terms are popularly and classically used in somewhat different senses.

Figurative or tropical language chiefly consists in the transfer of words to new senses, as by metaphor or metonymy. In the course of years, too, words change their meanings; and before the time of Dryden our whole vocabulary was much more fluid and adaptable than it has since become. Such authors as Bacon, Milton, and Sir Thomas Browne often used words derived from the Latin in some sense they originally had in Latin, though in English they had acquired another meaning. Spenser and Shakespeare, besides this practice, sometimes use words in a way that can only be justified by their choosing to have it so; whilst their contemporaries, Beaumont and Fletcher, write the perfect modern language, as Dryden observed. Lapse of time, however, is not the chief cause of variation in the sense of words. The matters which terms are used to denote are often so complicated or so refined in the assemblage, interfusion, or gradation of their qualities, that terms do not exist in sufficient abundance and discriminativeness to denote the things and, at the same time, to convey by connotation a determinate sense of their agreements and differences. In discussing politics, religion, ethics, æsthetics, this imperfection of language is continually felt; and the only escape from it, short of coining new words, is to use such words as we have, now in one sense, now in another somewhat different, and to trust to the context, or to the resources of the literary art, in order to convey the true meaning. Against this evil the having been born since Dryden is no protection. It behoves us, then, to remember that terms are not classifiable into Univocal and Æquivocal, but that all terms are susceptible of being used æquivocally, and that honesty and lucidity require us to try, as well as we can, to use each term univocally in the same context.

The context of any proposition always proceeds upon some assumption or understanding as to the scope of the discussion, which controls the interpretation of every statement and of every word. This was called by De Morgan the "universe of discourse": an older name for it, revived by Dr. Venn, and surely a better one, is *suppositio*. If we are talking of children, and 'play' is mentioned, the *suppositio* limits the suggestiveness of the word in one way; whilst if Monaco is the subject of conversation, the same word 'play,' under the influence of a different *suppositio*, excites altogether different ideas. Hence to ignore the *suppositio* is a great source of fallacies of equivocation. 'Man' is generally defined as a kind of animal; but 'animal' is often used as opposed to and excluding man. 'Liberal' has one meaning under the *suppositio* of politics, another with regard to culture, and still another as to the disposal of one's private means. Clearly, therefore, the connotation of general terms is relative to the *suppositio*, or "universe of discourse."

§ 5. Relative and Absolute Terms.—Some words go in couples or groups: like 'up-down,' 'former-latter,' 'father-mother-children,' 'hunter-prey,' 'cause-effect,' *etc.* These are called Relative Terms, and their nature, as explained by Mill, is that the connotations of the members of such a pair or group are derived from the same set of facts (the *fundamentum relationis*). There cannot be an 'up' without a 'down,' a 'father' without a 'mother' and 'child'; there cannot be a 'hunter' without something hunted, nor 'prey' without a pursuer. What makes a man a 'hunter' is his activities in pursuit; and what turns a chamois into 'prey' is its interest in these activities. The meaning of both terms, therefore, is derived from the same set of facts; neither term can be explained without explaining the other, because the relation between them is connoted by both; and neither can with propriety be used without reference to the other, or to some equivalent, as 'game' for 'prey.'

In contrast with such Relative Terms, others have been called Absolute or Non-relative. Whilst 'hunter' and 'prey' are relative, 'man' and 'chamois' have been considered absolute, as we may use them without thinking of any special connection between their meanings. However, if we believe in the unity of Nature and in the relativity of knowledge (that is, that all knowledge depends upon comparison, or a perception of the resemblances and differences of things), it follows that nothing can be completely understood except through its agreements or contrasts with everything else, and that all terms derive their connotation from the same set of facts, namely, from general experience. Thus both man and chamois are animals; this fact is an important part of the meaning of both terms, and to that extent they are relative terms. 'Five yards' and 'five minutes' are very different notions, yet they are profoundly related; for their very difference helps to make both notions distinct; and their intimate connection is shown in this, that five yards are traversed in a certain time, and that five minutes are measured by the motion of an index over some fraction of a yard upon the dial.

The distinction, then, between relative and non-relative terms must rest, not upon a fundamental difference between them (since, in fact, all words are relative), but upon the way in which words are used. We have seen that some words, such as 'up-down,' 'cause-effect,' can only be used relatively; and these may, for distinction, be called Correlatives. But other words, whose meanings are only partially interdependent, may often be used without attending to their relativity, and may then be considered as Absolute. We cannot say 'the hunter returned empty handed,' without implying that 'the prey escaped'; but we may say 'the man went supperless to bed,' without implying that 'the chamois rejoiced upon the mountain.' Such words as 'man' and 'chamois' may, then, in their use, be, as to one another, non-relative.

To illustrate further the relativity of terms, we may mention some of the chief classes of them.

Numerical order: 1st, 2nd, 3rd, *etc.*; 1st implies 2nd, and 2nd 1st; and 3rd implies 1st and 2nd, but these do not imply 3rd; and so on.

Order in Time or Place: before-after; early-punctual-late; right-middle-left; North-South, *etc.*

As to Extent, Volume, and Degree: greater-equal-less; large-medium-small; whole-part.

Genus and Species are a peculiar case of whole and part (*cf.* chaps. xxi.-ii.-iii.). Sometimes a term connotes all the attributes that another does, and more besides, which, as distinguishing it, are called differential. Thus 'man' connotes all that 'animal' does, and also (as *differentiæ*) the erect gait, articulate speech, and other attributes. In such a case as this, where there are well-marked classes, the term whose connotation is included in the others' is called a Genus of that Species. We have a Genus, triangle; and a Species, isosceles, marked off from all other triangles by the differential quality of having two equal sides: again—Genus, book; Species, quarto; Difference, having each sheet folded into four leaves.

There are other cases where these expressions 'genus' and 'species' cannot be so applied without a departure from usage, as, *e.g.*, if we call snow a species of the genus 'white,' for 'white' is not a recognised class. The connotation of white (*i.e.*, whiteness) is, however, part of the connotation of snow, just as the qualities of 'animal' are amongst those of 'man'; and for logical purposes it is desirable to use 'genus and species' to express that relativity of terms which consists in the connotation of one being part of the connotation of the other.

Two or more terms whose connotations severally include that of another term, whilst at the same time exceeding it, are (in relation to that other term) called Co-ordinate. Thus in relation to 'white,' snow and silver are co-ordinate; in relation to colour, yellow and red and blue are co-ordinate. And when all the terms thus related stand for recognised natural classes, the co-ordinate terms are called co-ordinate species; thus man and chamois are (in Logic) co-ordinate species of the genus animal.

§ 6. From such examples of terms whose connotations are related as whole and part, it is easy to see the general truth of the doctrine that as connotation decreases, denotation increases: for 'animal,' with less connotation than man or chamois, denotes many more objects; 'white,' with less connotation than snow or silver, denotes many more things, It is not, however, certain that this doctrine is always true in the concrete: since there may be a term connoting two or more qualities, all of which qualities are peculiar to all the things it denotes; and, if so, by subtracting one of the qualities from its connotation, we should not increase its denotation. If 'man,' for example, has among mammals the two peculiar attributes of erect gait and articulate speech, then, by omitting 'articulate speech' from the connotation of man, we could not apply the name to any more of the existing mammalia than we can at present. Still we might have been able to do so; there might have been an erect inarticulate ape, and perhaps there once was one; and, if so, to omit 'articulate' from the connotation of man would make the term 'man' denote that animal (supposing that there was no other difference to exclude it). Hence, potentially, an increase of the connotation of any term implies a decrease of its denotation. And, on the other hand, we can only increase the denotation of a term, or apply it to more objects, by decreasing its connotation; for, if the new things denoted by the term had already possessed its whole connotation, they must already have been denoted by it. However, we may increase the *known* denotation without decreasing the connotation, if we can discover the full connotation in things not formerly supposed to have it, as when dolphins were discovered to be mammals; or if we can impose the requisite qualities upon new individuals, as when by annexing some millions of Africans we extend the denotation of 'British subject' without altering its connotation.

Many of the things noticed in this chapter, especially in this section and the preceding, will be discussed at greater length in the chapters on Classification and Definition.

§ 7. Contradictory Relative Terms.—Every term has, or may have, another corresponding with it in such a way that, whatever differential qualities (§ 5) it connotes, this other connotes merely their absence; so that one or the other is always formally predicable of any Subject, but both these terms are never predicable of the same Subject in the same relation: such pairs of terms are called Contradictories. Whatever Subject we take, it is either visible or invisible, but not both; either human or non-human, but not both.

This at least is true formally, though in practice we should think ourselves trifled with if any one told us that 'A mountain is either human or non-human, but not both.' It is symbolic terms, such as X and x, that are properly said to be contradictories in relation to any subject whatever, S or M. For, as we have seen, the ordinary use of terms is limited by some *suppositio*, and this is true of Contradictories. 'Human' and 'non-human' may refer to zoological classification, or to the scope of physical, mental, or moral powers—as if we ask whether to flourish a dumbbell of a ton weight, or to know the future by intuition, or impeccability, be human or non-human. Similarly, 'visible' and 'invisible' refer either to the power of emitting or reflecting light, so that the words have no hold upon a sound or a scent, or else to power of vision and such qualifications as 'with the naked eye' or 'with a microscope.'

Again, the above definition of Contradictories tells us that they cannot be predicated of the same Subject "in the same relation"; that is, at the same time or place, or under the same conditions. The lamp is visible to me now, but will be invisible if I turn it out; one side of it is now visible, but the other is not: therefore without this restriction, "in the same relation," few or no terms would be contradictory.

If a man is called wise, it may mean 'on the whole' or 'in a certain action'; and clearly a man may for once be wise (or act wisely) who, on the whole, is not-wise. So that here again, by this ambiguity, terms that seem contradictory are predicable of the same subject, but not "in the same relation." In order to avoid the ambiguity, however, we have only to construct the term so as to express the relation, as 'wise on the whole'; and this immediately generates the contradictory 'not-wise on the whole.' Similarly, at one age a man may have black hair, at another not-black hair; but the difficulty is practically removable by stating the age referred to.

Still, this case easily leads us to a real difficulty in the use of contradictory terms, a difficulty arising from the continuous change or 'flux' of natural phenomena. If things are continually changing, it may be urged that contradictory terms are always applicable to the same subject, at least as fast as we can utter them: for if we have just said that a man's hair is black, since (like everything else) his hair is changing, it must now be not-black, though (to be sure) it may still seem black. The difficulty, such as it is, lies in this, that the human mind and its instrument language are not equal to the subtlety of Nature. All things flow, but the terms of human discourse assume a certain fixity of things; everything at every moment changes, but for the most part we can neither perceive this change nor express it in ordinary language.

This paradox, however, may, I suppose, be easily over-stated. The change that continually agitates Nature consists in the movements of masses or molecules, and such movements of things are compatible with a considerable persistence of their qualities. Not only are the molecular changes always going on in a piece of gold compatible with its remaining yellow, but its persistent yellowness depends on the continuance of some of those changes. Similarly, a man's hair may remain black for some years; though, no doubt, at a certain age its colour may begin to be problematical, and the applicability to it of 'black' or 'not-black' may become a matter of genuine anxiety. Whilst being on our guard, then, against fallacies of contradiction arising from the imperfect correspondence of fact with thought and language, we shall often have to put up with it. Candour and humility having been satisfied by the above acknowledgment of the subtlety of Nature, we may henceforward proceed upon the postulate—that it is possible to use contradictory terms such as cannot both be predicated of the same subject in the same relation, though one of them may be; that, for example, it may be truly said of a man for some years that his hair is black; and, if so, that during those years to call it not-black is false or extremely misleading.

The most opposed terms of the literary vocabulary, however, such as 'wise-foolish,' 'old-young,' 'sweet-bitter,' are rarely true contradictories: wise and foolish, indeed, cannot be predicated of the same man in the same relation; but there are many middling men, of whom neither can be predicated on the whole. For the comparison of quantities, again, we have three correlative terms, 'greater—equal—less,' and none of these is the contradictory of either of the others. In fact, the contradictory of any term is one that denotes the sum of its co-ordinates (§ 6); and to obtain a contradictory, the surest way is to coin one by prefixing to the given term the particle 'not' or (sometimes) 'non': as 'wise, not-wise,' 'human, non-human,' 'greater, not-greater.'

The separate word 'not' is surer to constitute a contradictory than the usual prefixes of negation, 'un-' or 'in-,' or even 'non'; since compounds of these are generally warped by common use from a purely negative meaning. Thus, 'Nonconformist' does not denote everybody who fails to conform. 'Unwise' is not equivalent to 'not-wise,' but means 'rather foolish'; a very foolish action is not-wise, but can only be called unwise by meiosis or irony. Still, negatives formed by 'in' or 'un' or 'non' are sometimes really contradictory of their positives; as 'visible, invisible,' 'equal, unequal.'

§ 8. The distinction between Positive and Negative terms is not of much value in Logic, what importance would else attach to it being absorbed by the more definite distinction of contradictories. For contradictories are positive and negative in essence and, when least ambiguously stated, also in form. And, on the other hand, as we have seen, when positive and negative terms are not contradictory, they are misleading. As with 'wise-unwise,' so with many others, such as 'happy-unhappy'; which are not contradictories; since a man may be neither happy nor unhappy, but indifferent, or (again) so miserable that he can only be called unhappy by a figure of speech. In fact, in the common vocabulary a formal negative often has a limited positive sense; and this is the case with unhappy, signifying the state of feeling in the milder shades of Purgatory.

When a Negative term is fully contradictory of its Positive it is said to be Infinite; because it denotes an unascertained multitude of things, a multitude only limited by the positive term and the *suppositio*; thus 'not-wise' denotes all except the wise, within the *suppositio* of 'intelligent beings.' Formally (disregarding any *suppositio*), such a negative term stands for all possible terms except its positive: x denotes everything but X; and 'not-wise' may be taken to include stones, triangles and hippogriffs. And even in this sense, a negative term has some positive meaning, though a very indefinite one, not a specific positive force like 'unwise' or 'unhappy': it denotes any and everything that has not the attributes connoted by the corresponding positive term.

Privative Terms connote the absence of a quality that normally belongs to the kind of thing denoted, as 'blind' or 'deaf.' We may predicate 'blind' or 'deaf' of a man, dog or cow that happens not to be able to see or hear, because the powers of seeing and hearing generally belong to those species; but of a stone or idol these terms can only be used figuratively. Indeed, since the contradictory of a privative carries with it the privative limitation, a stone is strictly 'not-blind': that is, it is 'not-something-that-normally-having-sight-wants-it.'

Contrary Terms are those that (within a certain genus or *suppositio*) severally connote differential qualities that are, in fact, mutually incompatible in the same relation to the same thing, and therefore cannot be predicated of the same subject in the same relation; and, so far, they resemble Contradictory Terms: but they differ from contradictory terms in this, that the differential quality connoted by each of them is definitely positive; no Contrary Term is infinite, but is limited to part of the *suppositio* excluded by the others; so that, possibly, neither of two Contraries is truly predicable of a given subject. Thus 'blue' and 'red' are Contraries, for they cannot both be predicated of the same thing in the same relation; but are not Contradictories, since, in a given case, neither may be predicable: if a flower is blue in a certain part, it cannot in the same part be red; but it may be neither blue nor red, but yellow; though it is certainly either blue or not-blue. All co-ordinate terms are formal Contraries; but if, in fact, a series of co-ordinates comprises only two (as male-female), they are empirical Contradictories; since each includes all that area of the *suppositio* which the other excludes.

The extremes of a series of co-ordinate terms are Opposites; as, in a list of colours, white and black, the most strongly contrasted, are said to be opposites, or as among moods of feeling, rapture and misery are opposites. But this distinction is of slight logical importance. Imperfect Positive and Negative couples, like 'happy and unhappy,' which (as we have seen) are not contradictories, are often called Opposites.

The members of any series of Contraries are all included by any one of them and its contradictory, as all colours come under 'red' and 'not-red,' all moods of feeling under 'happy' and 'not-happy.'

CHAPTER V

THE CLASSIFICATION OF PROPOSITIONS

§ 1. Logicians classify Propositions according to Quantity, Quality, Relation and Modality.

As to Quantity, propositions are either Universal or Particular; that is to say, the predicate is affirmed or denied either of the whole subject or of a part of it—of *All* or of *Some S*.

All S is P (that is, P is predicated of *all S*). *Some S is P* (that is, P is predicated of *some S*).

An Universal Proposition may have for its subject a singular term, a collective, a general term distributed, or an abstract term.

(1) A proposition having a singular term for its subject, as *The Queen has gone to France*, is called a Singular Proposition; and some Logicians regard this as a third species of proposition with respect to quantity, distinct from the Universal and Particular; but that is needless.

(2) A collective term may be the subject, as *The Black Watch is ordered to India*. In this case, as well as in singular propositions, a predication is made concerning the whole subject as a whole.

(3) The subject may be a general term taken in its full denotation, as *All apes are sagacious*; and in this case a Predication is made concerning the whole subject distributively; that is, of each and everything the subject stands for.

(4) Propositions whose subjects are abstract terms, though they may seem to be formally Singular, are really as to their meaning distributive Universals; since whatever is true of a quality is true of whatever thing has that quality so far as that quality is concerned. *Truth will prevail* means that *All true propositions are accepted at last* (by sheer force of being true, in spite of interests, prejudices, ignorance and indifference). To bear this in mind may make one cautious in the use of abstract terms.

In the above paragraphs a distinction is implied between Singular and Distributive Universals; but, technically, every term, whether subject or predicate, when taken in its full denotation (or universally), is said to be 'distributed,' although this word, in its ordinary sense, would be directly applicable only to general terms. In the above examples, then, 'Queen,' 'Black Watch,' 'apes,' and 'truth' are all distributed terms. Indeed, a simple definition of the Universal Proposition is 'one whose subject is distributed.'

A Particular Proposition is one that has a general term for its subject, whilst its predicate is not affirmed or denied of everything the subject denotes; in other words, it is one whose subject is not distributed: as *Some lions inhabit Africa*.

In ordinary discourse it is not always explicitly stated whether predication is universal or particular; it would be very natural to say *Lions inhabit Africa*, leaving it, as far as the words go, uncertain whether we mean *all* or *some* lions. Propositions whose quantity is thus left indefinite are technically called 'preindesignate,' their quantity not being stated or designated by any introductory expression; whilst propositions whose quantity is expressed, as *All foundling-hospitals have a high death-rate*, or *Some wine is made from grapes*, are said to be 'predesignate.' Now, the rule is that preindesignate propositions are, for logical purposes, to be treated as particular; since it is an obvious precaution of the science of proof, in any practical application, *not to go beyond the evidence*. Still, the rule may be relaxed if the universal quantity of a preindesignate proposition is well known or admitted, as in *Planets shine with reflected light*—understood of the planets of our solar system at the present time. Again, such a proposition as *Man is the paragon of animals* is not a preindesignate, but an abstract proposition; the subject being elliptical for *Man according to his proper nature*; and the translation of it into a predesignate proposition is not *All men are paragons*; nor can *Some men* be sufficient, since an abstract can only be adequately rendered by a distributed term; but we must say, *All men who approach the ideal*. Universal real propositions, true without qualification, are very scarce; and we often substitute for them *general* propositions, saying perhaps—*generally, though not universally, S is P*. Such general propositions are, in strictness, particular; and the logical rules concerning universals cannot be applied to them without careful scrutiny of the facts.

The marks or predesignations of Quantity commonly used in Logic are: for Universals, *All, Any, Every, Whatever* (in the negative *No* or *No one*, see next §); for Particulars, *Some*.

Now *Some*, technically used, does not mean *Some only*, but *Some at least* (it may be one, or more, or all). If it meant '*Some only*,' every particular proposition would be an exclusive exponible (chap. ii. § 3); since *Only some men are wise* implies that *Some men are not wise*. Besides, it may often happen in an investigation that all the instances we have observed come under a certain rule, though we do not yet feel justified in regarding the rule as universal; and this situation is exactly met by the expression *Some* (*it may be all*).

The words *Many, Most, Few* are generally interpreted to mean *Some*; but as *Most* signifies that exceptions are known, and *Few* that the exceptions are the more numerous, propositions thus predesignate are in fact exponibles, mounting to *Some are* and *Some are not*. If to work with both forms be too cumbrous, so that we must choose one, apparently *Few are* should be treated as *Some are not*. The scientific course to adopt with propositions predesignate by *Most* or *Few*, is to collect statistics and determine the percentage; thus, *Few men are wise*—say 2 per cent.

The Quantity of a proposition, then, is usually determined entirely by the quantity of the subject, whether *all* or *some*. Still, the quantity of the predicate is often an important consideration; and though in ordinary usage the predicate is seldom predesignate, Logicians agree that in every Negative Proposition (see § 2) the predicate is 'distributed,' that is to say, is denied altogether of the subject, and that this is involved in the form of denial. To say *Some men are not brave*, is to declare that the quality for which men may be called brave is not found in any of the *Some men* referred to: and to say *No men are proof against flattery*, cuts off the being 'proof against flattery' entirely from the list of human attributes. On the other hand, every Affirmative Proposition is regarded as having an undistributed predicate; that is to say, its predicate is not affirmed exclusively of the subject. *Some men are wise* does not mean that 'wise' cannot be predicated of any other beings; it is equivalent to *Some men are wise* (whoever else may be). And *All elephants are sagacious* does not limit sagacity to elephants: regarding 'sagacious' as possibly denoting many animals of many species that exhibit the quality, this proposition is equivalent to '*All elephants are* some *sagacious animals*.' The affirmative predication of a quality does not imply exclusive possession of it as denial implies its complete absence; and, therefore, to regard the predicate of an affirmative proposition as distributed would be to go beyond the evidence and to take for granted what had never been alleged.

Some Logicians, seeing that the quantity of predicates, though not distinctly expressed, is recognised, and holding that it is the part of Logic "to make explicit in language whatever is implicit in thought," have proposed to exhibit the quantity of predicates by predesignation, thus: 'Some men are *some* wise (beings)'; 'some men are not *any* brave (beings)'; *etc.* This is called the Quantification of the Predicate, and leads to some modifications of Deductive Logic which will be referred to hereafter. (See § 5; chap. vii. § 4, and chap. viii. § 3.)

§ 2. As to Quality, Propositions are either Affirmative or Negative. An Affirmative Proposition is, formally, one whose copula is affirmative (or, has no negative sign), as *S—is—P, All men—are—partial to themselves*. A Negative Proposition is one whose copula is negative (or, has a negative sign), as *S—is not—P, Some men—are not—proof against flattery*. When, indeed, a Negative Proposition is of Universal Quantity, it is stated thus: *No S is P, No men are proof against flattery*; but, in this case, the detachment of the negative sign from the copula and its association with the subject is merely an accident of our idiom; the proposition is the same as *All men—are not—proof against flattery*. It must be distinguished, therefore, from such an expression as *Not every man is proof against flattery*; for here the negative sign really restricts the subject; so that the meaning is—*Some men at most* (it may be *none*) *are proof against flattery*; and thus the proposition is Particular, and is rendered—*Some men—are not—proof against flattery*.

When the negative sign is associated with the predicate, so as to make this an Infinite Term (chap. iv. § 8), the proposition is called an Infinite Proposition, as *S is not-P* (or *p*), *All men are—incapable of resisting flattery*, or *are—not-proof against flattery*.

Infinite propositions, when the copula is affirmative, are formally, themselves affirmative, although their force is chiefly negative; for, as the last example shows, the difference between an infinite and a negative proposition may depend upon a hyphen. It has been proposed, indeed, with a view to superficial simplification, to turn all Negatives into Infinites, and thus render all propositions Affirmative in Quality. But although every proposition both affirms and denies something according to the aspect in which you regard it (as *Snow is white* denies that it is any other colour, and *Snow is not blue* affirms that it is some other colour), yet there is a great difference between the definite affirmation of a genuine affirmative and the vague affirmation of a negative or infinite; so that materially an affirmative infinite is the same as a negative.

Generally Mill's remark is true, that affirmation and denial stand for distinctions of fact that cannot be got rid of by manipulation of words. Whether granite sinks in water, or not; whether the rook lives a hundred years, or not; whether a man has a hundred dollars in his pocket, or not; whether human bones have ever been found in Pliocene strata, or not; such alternatives require distinct forms of expression. At the same time, it may be granted that many facts admit of being stated with nearly equal propriety in either Quality, as *No man is proof against flattery*, or *All men are open to flattery*.

But whatever advantage there is in occasionally changing the Quality of a proposition may be gained by the process of Obversion (chap. vii. § 5); whilst to use only one Quality would impair the elasticity of logical expression. It is a postulate of Logic that the negative sign may be transferred from the copula to the predicate, or from the predicate to the copula, without altering the sense of a proposition; and this is justified by the experience that not to have an attribute and to be without it are the same thing.

§ 3. A. I. E. O.—Combining the two kinds of Quantity, Universal and Particular, with the two kinds of Quality, Affirmative and Negative, we get four simple types of proposition, which it is usual to symbolise by the letters A. I. E. O., thus:

A   Universal   — All S is P.

A.	Affirmative	
I.	Particular Affirmative	— Some S is P.
E.	Universal Negative	— No S is P.
O.	Particular Negative	— Some S is not P.

As an aid to the remembering of these symbols we may observe that A. and I. are the first two vowels in *affirmo* and that E. and O. are the vowels in *nego*.

It must be acknowledged that these four kinds of proposition recognised by Formal Logic constitute a very meagre selection from the list of propositions actually used in judgment and reasoning.

Those Logicians who explicitly quantify the predicate obtain, in all, eight forms of proposition according to Quantity and Quality:

U.	Toto-total Affirmative	— All X is all Y.
A.	Toto-partial Affirmative	— All X is some Y.
Y.	Parti-total Affirmative	— Some X is all Y.
I.	Parti-partial Affirmative	— Some X is some Y.
E.	Toto-total Negative	— No X is any Y.
η.	Toto-partial Negative	— No X is some Y.
O.	Parti-total Negative	— Some X is not any Y.
ω.	Parti-partial Negative	— Some X is not some Y.

Here A. I. E. O. correspond with those similarly symbolised in the usual list, merely designating in the predicates the quantity which was formerly treated as implicit.

§ 4. As to Relation, propositions are either Categorical or Conditional. A Categorical Proposition is one in which the predicate is directly affirmed or denied of the subject without any limitation of time, place, or circumstance, extraneous to the subject, as *All men in England are secure of justice*; in which proposition, though there is a limitation of place ('in England'), it is included in the subject. Of this kind are nearly all the examples that have yet been given, according to the form *S is P*.

A Conditional Proposition is so called because the predication is made under some limitation or condition not included in the subject, as *If a man live in England, he is secure of justice*. Here the limitation 'living in England' is put into a conditional sentence extraneous to the subject, 'he,' representing any man.

Conditional propositions, again, are of two kinds—Hypothetical and Disjunctive. Hypothetical propositions are those that are limited by an explicit conditional sentence, as above, or thus: *If Joe Smith was a prophet, his followers have been unjustly persecuted*. Or in symbols thus:

If A is, B is; If A is B, A is C; If A is B, C is D.

Disjunctive propositions are those in which the condition under which predication is made is not explicit but only implied under the disguise of an alternative proposition, as *Joe Smith was either a prophet or an impostor*. Here there is no direct predication concerning Joe Smith, but only a predication of one of the alternatives conditionally on the other being denied, as, *If Joe Smith was not a prophet he was an impostor*; or, *If he was not an impostor, he was a prophet*. Symbolically, Disjunctives may be represented thus:

A is either B or C, Either A is B or C is D.

Formally, every Conditional may be expressed as a Categorical. For our last example shows how a Disjunctive may be reduced to two Hypotheticals (of which one is redundant, being the contrapositive of the other; see chap. vii. § 10). And a Hypothetical is reducible to a Categorical thus: *If the sky is clear, the night is cold* may be read—*The case of the sky being clear is a case of the night being cold*; and this, though a clumsy plan, is sometimes convenient. It would be better to say *The sky being clear is a sign of the night being cold*, or a condition of it. For, as Mill says, the essence of a Hypothetical is to state that one clause of it (the indicative) may be inferred from the other (the conditional). Similarly, we might write: *Proof of Joe Smith's not being a prophet is a proof of his being an impostor*.

This turning of Conditionals into Categoricals is called a Change of Relation; and the process may be reversed: *All the wise are virtuous* may be written, *If any man is wise he is virtuous*; or, again, *Either a man is not-wise or he is virtuous*. But the categorical form is usually the simplest.

If, then, as substitutes for the corresponding conditionals, categoricals are formally adequate, though sometimes inelegant, it may be urged that Logic has nothing to do with elegance; or that, at any rate, the chief elegance of science is economy, and that therefore, for scientific purposes, whatever we may write further about conditionals must be an ugly excrescence. The scientific purpose of Logic is to assign the conditions of proof. Can we, then, in the conditional form prove anything that cannot be proved in the categorical? Or does a conditional require to be itself proved by any method not applicable to the Categorical? If not, why go on with the discussion of Conditionals? For all laws of Nature, however stated, are essentially categorical. 'If a straight line falls on another straight line, the adjacent angles are together equal to two right angles'; 'If a body is unsupported, it falls'; 'If population increases, rents tend to rise': here 'if' means 'whenever' or 'all cases in which'; for to raise a doubt whether a straight line is ever conceived to fall upon another, whether bodies are ever unsupported, or population ever increases, is a superfluity of scepticism; and plainly the hypothetical form has nothing to do with the proof of such propositions, nor with inference from them.

Still, the disjunctive form is necessary in setting out the relation of contradictory terms, and in stating a Division (chap. xxi.), whether formal (*as A is B or not-B*) or material (as *Cats are white, or black, or tortoiseshell, or tabby*). And in some cases the hypothetical form is useful. One of these occurs where it is important to draw attention to the condition, as something doubtful or especially requiring examination. *If there is a resisting medium in space, the earth will fall into the sun; If the Corn Laws are to be re-enacted, we had better sell railways and buy land*: here the hypothetical form draws attention to the questions whether there is a resisting medium in space, whether the Corn Laws are likely to be re-enacted; but as to methods of inference and proof, the hypothetical form has nothing to do with them. The propositions predicate causation: *A resisting medium in space is a condition of the earth's falling into the sun; A Corn Law is a condition of the rise of rents, and of the fall of railway profits.*

A second case in which the hypothetical is a specially appropriate form of statement occurs where a proposition relates to a particular matter and to future time, as *If there be a storm to-morrow, we shall miss our picnic*. Such cases are of very slight logical interest. It is as exercises in formal thinking that hypotheticals are of most value; inasmuch as many people find them more difficult than categoricals to manipulate.

In discussing Conditional Propositions, the conditional sentence of a Hypothetical, or the first alternative of a Disjunctive, is called the Antecedent; the indicative sentence of a Hypothetical, or the second alternative of a Disjunctive, is called the Consequent.

Hypotheticals, like Categoricals, have been classed according to Quantity and Quality. Premising that the quantity of a Hypothetical depends on the quantity of its Antecedent (which determines its limitation), whilst its quality depends on the quality of its consequent (which makes the predication), we may exhibit four forms:

A. *If A is B, C is D*; I. *Sometimes when A is B, C is D*; E. *If A is B, C is not D*; O. *Sometimes when A is B, C is not D*.

But I. and O. are rarely used.

As for Disjunctives, it is easy to distinguish the two quantities thus:

A. *Either A is B, or C is D*; I. *Sometimes either A is B or C is D.*

But I. is rarely used. The distinction of quality, however, cannot be made: there are no true negative forms; for if we write—

Neither is A B, nor C D,

there is here no alternative predication, but only an Exponible equivalent to *No A is B, and No C is D*. And if we write—

Either A is not B, or C is not D,

this is affirmative as to the alternation, and is for all methods of treatment equivalent to A.

Logicians are divided in opinion as to the interpretation of the conjunction 'either, or'; some holding that it means 'not both,' others that it means 'it may be both.' Grammatical usage, upon which the question is sometimes argued, does not seem to be established in favour of either view. If we say *A man so precise in his walk and conversation is either a saint or a consummate hypocrite*; or, again, *One who is happy in a solitary life is either more or less than man*; we cannot in such cases mean that the subject may be both. On the other hand, if it be said that *the author of 'A Tale of a Tub' is either a misanthrope or a dyspeptic*, the alternatives are not incompatible. Or, again, given that *X. is a lunatic, or a lover, or a poet*, the three predicates have much congruity.

It has been urged that in Logic, language should be made as exact and definite as possible, and that this requires the exclusive interpretation 'not both.' But it seems a better argument, that Logic (1) should be able to express all meanings, and (2), as the science of evidence, must not assume more than is given; to be on the safe side, it must in doubtful cases assume the least, just as it generally assumes a preindesignate term to be of particular quantity; and, therefore 'either, or' means 'one, or the other, or both.'

However, when both the alternative propositions have the same subject, as *Either A is B, or A is C*, if the two predicates are contrary or contradictory terms (as 'saint' and 'hypocrite,' or 'saint' and 'not-saint'), they cannot in their nature be predicable in the same way of the same subject; and, therefore, in such a case 'either, or' means one or the other, but not both in the same relation. Hence it seems necessary to admit that the conjunction 'either, or' may sometimes require one interpretation, sometimes the other; and the rule is that it implies the further possibility 'or both,' except when both alternatives have the same subject whilst the predicates are contrary or contradictory terms.

If, then, the disjunctive *A is either B or C* (*B* and *C* being contraries) implies that both alternatives cannot be true, it can only be adequately rendered in hypotheticals by the two forms—(1) *If A is B, it is not C*, and (2)*If A is not B, it is C*. But if the disjunctive *A is either B or C* (*B* and *C* not being contraries) implies that both may be true, it will be adequately translated into a hypothetical by the single form, *If A is not B, it is C*. We cannot translate it into—*If A is B, it is not C*, for, by our supposition, if '*A is B*' is true, it does not follow that '*A is C*' must be false.

Logicians are also divided in opinion as to the function of the hypothetical form. Some think it expresses doubt; for the consequent depends on the antecedent, and the antecedent, introduced by 'if,' may or may not be realised, as in *If the sky is clear, the night is cold*: whether the sky is, or is not, clear being supposed to be uncertain. And we have seen that some hypothetical propositions seem designed to draw attention to such uncertainty, as—*If there is a resisting medium in space, etc.* But other Logicians lay stress upon the connection of the clauses as the important matter: the statement is, they say, that the consequent may be inferred from the antecedent. Some even declare that it is given as a necessary inference; and on this ground Sigwart rejects particular hypotheticals, such as *Sometimes when A is B, C is D*; for if it happens only sometimes the connexion cannot be necessary. Indeed, it cannot even be probably inferred without further grounds. But this is also true whenever the antecedent and consequent are concerned with different matter. For example, *If the soul is simple, it is indestructible*. How do you know that? Because *Every simple substance is indestructible*. Without this further ground there can be no inference. The fact is that conditional forms often cover assertions that are not true complex propositions but a sort of euthymemes (chap. xi. § 2), arguments abbreviated and rhetorically disguised. Thus: *If patience is a virtue there are painful virtues*—an example from Dr. Keynes. Expanding this we have—

Patience is painful; Patience is a virtue: ∴ Some virtue is painful.

And then we see the equivocation of the inference; for though patience be painful *to learn*, it is not painful *as a virtue* to the patient man.

The hypothetical, '*If Plato was not mistaken poets are dangerous citizens*,' may be considered as an argument against the laureateship, and may be expanded (informally) thus: 'All Plato's opinions deserve respect; one of them was that poets are bad citizens; therefore it behoves us to be chary of encouraging poetry.' Or take this disjunctive, '*Either Bacon wrote the works ascribed to Shakespeare, or there were two men of the highest genius in the same age and country*.' This means that it is not likely there should be two such men, that we are sure of Bacon, and therefore ought to give him all the glory. Now, if it is the part of Logic 'to make explicit in language all that is implicit in thought,' or to put arguments into the form in which they can best be examined, such propositions as the above ought to be analysed in the way suggested, and confirmed or refuted according to their real intention.

We may conclude that no single function can be assigned to all hypothetical propositions: each must be treated according to its own meaning in its own context.

§ 5. As to Modality, propositions are divided into Pure and Modal. A Modal proposition is one in which the predicate is affirmed or denied, not simply but *cum modo*, with a qualification. And some Logicians have considered any adverb occurring in the predicate, or any sign of past or future tense, enough to constitute a modal: as 'Petroleum is *dangerously* inflammable'; 'English *will be* the universal language.' But far the most important kind of modality, and the only one we need consider, is that which is signified by some qualification of the predicate as to the degree of certainty with which it is affirmed or denied. Thus, 'The bite of the cobra is *probably* mortal,' is called a Contingent or Problematic Modal: 'Water is *certainly* composed of oxygen and hydrogen' is an Assertory or Certain Modal: 'Two straight lines *cannot* enclose a space' is a Necessary or Apodeictic Modal (the opposite being inconceivable). Propositions not thus qualified are called Pure.

Modal propositions have had a long and eventful history, but they have not been found tractable by the resources of ordinary Logic, and are now generally neglected by the authors of text-books. No doubt such propositions are the commonest in ordinary discourse, and in some rough way we combine them and draw inferences from them. It is understood that a combination of assertory or of apodeictic premises may warrant an assertory or an apodeictic conclusion; but that if we combine either of these with a problematic premise our conclusion becomes problematic; whilst the combination of two problematic premises gives a conclusion less certain than either. But if we ask 'How much less certain?' there is no answer. That the modality of a conclusion follows the less certain of the premises combined, is inadequate for scientific guidance; so that, as Deductive Logic can get no farther than this, it has abandoned the discussion of Modals. To endeavour to determine the degree of certainty attaching to a problematic judgment is not, however, beyond the reach of Induction, by analysing circumstantial evidence, or by collecting statistics with regard to it. Thus, instead of 'The cobra's bite is *probably* fatal,' we might find that it is fatal 80 times in 100. Then, if we know that of those who go to India 3 in 1000 are bitten, we can calculate what the chances are that any one going to India will die of a cobra's bite (chap. xx.).

§ 6. Verbal and Real Propositions.—Another important division of propositions turns upon the relation of the predicate to the subject in respect of their connotations. We saw, when discussing Relative Terms, that the connotation of one term often implies that of another; sometimes reciprocally, like 'master' and 'slave'; or by inclusion, like species and genus; or by exclusion, like contraries and contradictories. When terms so related appear as subject and predicate of the same proposition, the result is often tautology—*e.g., The master has authority over his slave; A horse is an animal; Red is not blue; British is not foreign*. Whoever knows the meaning of 'master,' 'horse,' 'red,' 'British,' learns nothing from these propositions. Hence they are called Verbal propositions, as only expounding the sense of words, or as if they were propositions only by satisfying the forms of language, not by fulfilling the function of propositions in conveying a knowledge of facts. They are also called 'Analytic' and 'Explicative,' when they separate and disengage the elements of the connotation of the subject. Doubtless, such propositions may be useful to one who does not know the language; and Definitions, which are verbal propositions whose predicates analyse the whole connotations of their subjects, are indispensable instruments of science (see chap. xxii.).

Of course, hypothetical propositions may also be verbal, as *If the soul be material it is extended*; for 'extension' is connoted by 'matter'; and, therefore, the corresponding disjunctive is verbal—*Either the soul is not material, or it is extended*. But a true divisional disjunctive can never be verbal (chap. xxi. § 4, rule 1).

On the other hand, when there is no such direct relation between subject and predicate that their connotations imply one another, but the predicate connotes something that cannot be learnt from the connotation of the subject, there is no longer tautology, but an enlargement of meaning—*e.g., Masters are degraded by their slaves; The horse is the noblest animal; Red is the favourite colour of the British army; If the soul is simple, it is indestructible*. Such propositions are called Real, Synthetic, or Ampliative, because they are propositions for which a mere understanding of their subjects would be no substitute, since the predicate adds a meaning of its own concerning matter of fact.

To any one who understands the language, a verbal proposition can never be an inference or conclusion from evidence; nor can a verbal proposition ever furnish grounds for an inference, except as to the meaning of words. The subject of real and verbal propositions will inevitably recur in the chapters on Definition; but tautologies are such common blemishes in composition, and such frequent pitfalls in argument, that attention cannot be drawn to them too early or too often.

CHAPTER VI
CONDITIONS OF IMMEDIATE INFERENCE

§ 1. The word Inference is used in two different senses, which are often confused but should be carefully distinguished. In the first sense, it means a process of thought or reasoning by which the mind passes from facts or statements presented, to some opinion or expectation. The data may be very vague and slight, prompting no more than a guess or surmise; as when we look up at the sky and form some expectation about the weather, or from the trick of a man's face entertain some prejudice as to his character. Or the data may be important and strongly significant, like the footprint that frightened Crusoe into thinking of cannibals, or as when news of war makes the city expect that Consols will fall. These are examples of the act of inferring, or of inference as a process; and with inference in this sense Logic has nothing to do; it belongs to Psychology to explain how it is that our minds pass from one perception or thought to another thought, and how we come to conjecture, conclude and believe (cf. chap. i. § 6).

In the second sense, 'inference' means not this process of guessing or opining, but the result of it; the surmise, opinion, or belief when formed; in a word, the conclusion: and it is in this sense that Inference is treated of in Logic. The subject-matter of Logic is an inference, judgment or conclusion concerning facts, embodied in a proposition, which is to be examined in relation to the evidence that may be adduced for it, in order to determine whether, or how far, the evidence amounts to proof. Logic is the science of Reasoning in the sense in which 'reasoning' means giving reasons, for it shows what sort of reasons are good. Whilst Psychology explains how the mind goes forward from data to conclusions, Logic takes a conclusion and goes back to the data, inquiring whether those data, together with any other evidence (facts or principles) that can be collected, are of a nature to warrant the conclusion. If we think that the night will be stormy, that John Doe is of an amiable disposition, that water expands in freezing, or that one means to national prosperity is popular education, and wish to know whether we have evidence sufficient to justify us in holding these opinions, Logic can tell us what form the evidence should assume in order to be conclusive. What *form* the evidence should assume: Logic cannot tell us what kinds of fact are proper evidence in any of these cases; that is a question for the man of special experience in life, or in science, or in business. But whatever facts constitute the evidence, they must, in order to prove the point, admit of being stated in conformity with certain principles or conditions; and of these principles or conditions Logic is the science. It deals, then, not with the subjective process of inferring, but with the objective grounds that justify or discredit the inference.

§ 2. Inferences, in the Logical sense, are divided into two great classes, the Immediate and the Mediate, according to the character of the evidence offered in proof of them. Strictly, to speak of inferences, in the sense of conclusions, as immediate or mediate, is an abuse of language, derived from times before the distinction between inference as process and inference as result was generally felt. No doubt we ought rather to speak of Immediate and Mediate Evidence; but it is of little use to attempt to alter the traditional expressions of the science.

An Immediate Inference, then, is one that depends for its proof upon only one other proposition, which has the same, or more extensive, terms (or matter). Thus that *one means to national prosperity is popular education* is an immediate inference, if the evidence for it is no more than the admission that *popular education is a means to national prosperity:* Similarly, it is an immediate inference that *Some authors are vain*, if it be granted that *All authors are vain*.

An Immediate Inference may seem to be little else than a verbal transformation; some Logicians dispute its claims to be called an inference at all, on the ground that it is identical with the pretended evidence. If we attend to the meaning, say they, an immediate inference does not really express any new judgment; the fact expressed by it is either the same as its evidence, or is even less significant. If from *No men are gods* we prove that *No gods are men*, this is nugatory; if we prove from it that *Some men are not gods*, this is to emasculate the sense, to waste valuable information, to lose the commanding sweep of our universal proposition.

Still, in Logic, it is often found that an immediate inference expresses our knowledge in a more convenient form than that of the evidentiary proposition, as will appear in the chapter on Syllogisms and elsewhere. And by transforming an universal into a particular proposition, as *No men are gods*, therefore, *Some men are not gods*,—we get a statement which, though weaker, is far more easily proved; since a single instance suffices. Moreover, by drawing all possible immediate inferences from a given proposition, we see it in all its aspects, and learn all that is implied in it.

A Mediate Inference, on the other hand, depends for its evidence upon a plurality of other propositions (two or more) which are connected together on logical principles. If we argue—

No men are gods; Alexander the Great is a man; ∴ Alexander the Great is not a god:

this is a Mediate Inference. The evidence consists of two propositions connected by the term 'man,' which is common to both (a Middle Term), mediating between 'gods' and 'Alexander.' Mediate Inferences comprise Syllogisms with their developments, and Inductions; and to discuss them further at present would be to anticipate future chapters. We must now deal with the principles or conditions on which Immediate Inferences are valid: commonly called the "Laws of Thought."

§ 3. The Laws of Thought are conditions of the logical statement and criticism of all sorts of evidence; but as to Immediate Inference, they may be regarded as the only conditions it need satisfy. They are often expressed thus: (1) The principle of Identity—'*Whatever is, is*'; (2) The principle of Contradiction—'*It is impossible for the same thing to be and not be*'; (3) The principle of Excluded Middle—'*Anything must either be or not be.*' These principles are manifestly not 'laws' of thought in the sense in which 'law' is used in Psychology; they do not profess to describe the actual mental processes that take place in judgment or reasoning, as the 'laws of association of ideas' account for memory and recollection. They are not natural laws of thought; but, in relation to thought, can only be regarded as laws when stated as precepts, the observance of which (consciously or not) is necessary to clear and consistent thinking: *e.g.*, Never assume that the same thing can both be and not be.

However, treating Logic as the science of thought only as embodied in propositions, in respect of which evidence is to be adduced, or which are to be used as evidence of other propositions, the above laws or principles must be restated as the conditions of consistent argument in such terms as to be directly applicable to propositions. It was shown in the chapter on the connotation of terms, that terms are assumed by Logicians to be capable of definite meaning, and of being used univocally in the same context; if, or in so far as, this is not the case, we cannot understand one another's reasons nor even pursue in solitary meditation any coherent train of argument. We saw, too, that the meanings of terms were related to one another: some being full correlatives; others partially inclusive one of another, as species of genus; others mutually incompatible, as contraries; or alternatively predicable, as contradictories. We now assume that propositions are capable of definite meaning according to the meaning of their component terms and of the relation between them; that the meaning, the fact asserted or denied, is what we are really concerned to prove or disprove; that a mere change in the words that constitute our terms, or of construction, does not affect the truth of a proposition as long as the meaning is not altered, or (rather) as long as no fresh meaning is introduced; and that if the meaning of any proposition is true, any other proposition that denies it is false. This postulate is plainly necessary to consistency of statement and discourse; and consistency is necessary, if our thought or speech is to correspond with the unity and coherence of Nature and experience; and the Laws of Thought or Conditions of Immediate Inference are an analysis of this postulate.

§ 4. The principle of Identity is usually written symbolically thus: *A is A; not-A is not-A.* It assumes that there is something that may be represented by a term; and it requires that, in any discussion, *every relevant term, once used in a definite sense, shall keep that meaning throughout*. Socrates in his father's workshop, at the battle of Delium, and in prison, is assumed to be the same man denotable by the same name; and similarly, 'elephant,' or 'justice,' or 'fairy,' in the same context, is to be understood of the same thing under the same *suppositio*.

But, further, it is assumed that of a given term another term may be predicated again and again in the same sense under the same conditions; that is, we may speak of the identity of meaning in a proposition as well as in a term. To symbolise this we ought to alter the usual formula for Identity and write it thus: *If B is A, B is A; if B is not-A, B is not-A.* If Socrates is wise, he is wise; if fairies frequent the moonlight, they do; if Justice is not of this world, it is not. *Whatever affirmation or denial we make concerning any subject, we are bound to adhere to it for the purposes of the current argument or investigation.* Of course, if our assertion turns out to be false, we must not adhere to it; but then we must repudiate all that we formerly deduced from it.

Again, *whatever is true or false in one form of words is true or false in any other*: this is undeniable, for the important thing is identity of meaning; but in Formal Logic it is not very convenient. If Socrates is wise, is it an identity to say 'Therefore the master of Plato is wise'; or, further that he 'takes enlightened views of life'? If *Every man is fallible*, is it an identical proposition that *Every man is liable to error*? It seems pedantic to demand a separate proposition that *Fallible is liable to error*. But, on the other hand, the insidious substitution of one term for another speciously identical, is a chief occasion of fallacy. How if we go on to argue: therefore, *Every man is apt to blunder, prone to confusion of thought, inured to self-contradiction*? Practically, the substitution of identities must be left to candour and good-sense; and may they increase among us. Formal Logic is, no doubt, safest with symbols; should, perhaps, content itself with A and B; or, at least, hardly venture beyond Y and Z.

§ 5. The principle of Contradiction is usually written symbolically, thus: *A is not not-A*. But, since this formula seems to be adapted to a single term, whereas we want one that is applicable to propositions, it may be better to write it thus: *B is not both A and not-A*. That is to say: *if any term may be affirmed of a subject, the contradictory term may, in the same relation, be denied of it*. A leaf that is green on one side of it may be not-green on the other; but it is not both green and not-green on the same surface, at the same time, and in the same light. If a stick is straight, it is false that it is at the same time not-straight: having granted that two angles are equal, we must deny that they are unequal.

But is it necessarily false that the stick is 'crooked'; must we deny that either angle is 'greater or less' than the other? How far is it permissible to substitute any other term for the formal contradictory? Clearly, the principle of Contradiction takes for granted the principle of Identity, and is subject to the same difficulties in its practical application. As a matter of fact and common sense, if we affirm any term of a Subject, we are bound to deny of that Subject, in the same relation, not only the contradictory but all synonyms for this, and also all contraries and opposites; which, of course, are included in the contradictory. But who shall determine what these are? Without an authoritative Logical Dictionary to refer to, where all contradictories, synonyms, and contraries may be found on record, Formal Logic will hardly sanction the free play of common sense.

The principle of Excluded Middle may be written: *B is either A or not-A*; that is, *if any term be denied of a subject, the contradictory term may, in the same relation, be affirmed*. Of course, we may deny that a leaf is green on one side without being bound to affirm that it is not-green on the other. But in the same relation a leaf is either green or not-green; at the same time, a stick is either bent or not-bent. If we deny that A is greater than B, we must affirm that it is not-greater than B.

Whilst, then, the principle of Contradiction (that 'of contradictory predicates, one being affirmed, the other is denied') might seem to leave open a third or middle course, the denying of both contradictories, the principle of Excluded Middle derives its name from the excluding of this middle course, by declaring that the one or the other must be affirmed. Hence the principle of Excluded Middle does not hold good of mere contrary terms. If we deny that a leaf is green, we are not bound to affirm it to be yellow; for it may be red; and then we may deny both contraries, yellow and green. In fact, two contraries do not between them cover the whole predicable area, but contradictories do: the form of their expression is such that (within the *suppositio*) each includes all that the other excludes; so that the subject (if brought within the *suppositio*) must fall under the one or the other. It may seem absurd to say that Mont Blanc is either wise or not-wise; but how comes any mind so ill-organised as to introduce Mont Blanc into this strange company? Being there, however, the principle is inexorable: Mont Blanc is not-wise.

In fact, the principles of Contradiction and Excluded Middle are inseparable; they are implicit in all distinct experience, and may be regarded as indicating the two aspects of Negation. The principle of Contradiction says: *B is not both A and not-A*, as if *not-A* might be nothing at all; this is abstract negation. But the principle of Excluded Middle says: *Granting that B is not A, it is still something*—namely, *not-A*; thus bringing us back to the concrete experience of a continuum in which the absence of one thing implies the presence of something else. Symbolically: to deny that B is A is to affirm that B is not A, and this only differs by a hyphen from B is not-A.

These principles, which were necessarily to some extent anticipated in [chap. iv. § 7](), the next chapter will further illustrate.

§ 6. But first we must draw attention to a maxim (also already mentioned), which is strictly applicable to Immediate Inferences, though (as we shall see) in other kinds of proof it may be only a formal condition: this is the general caution *not to go beyond the evidence*. An immediate inference ought to contain nothing that is not contained (or formally implied) in the proposition by which it is proved. With respect to quantity in denotation, this caution is embodied in the rule 'not to distribute any term that is not given distributed.' Thus, if there is a predication concerning 'Some S,' or 'Some men,' as in the forms I. and O., we cannot infer anything concerning 'All S.' or 'All men'; and, as we have seen, if a term is given us preindesignate, we are generally to take it as of particular quantity. Similarly, in the case of affirmative propositions, we saw that this rule requires us to assume that their predicates are undistributed.

As to the grounds of this maxim, not to go beyond the evidence, not to distribute a term that is given as undistributed, it is one of the things so plain that to try to justify is only to obscure them. Still, we must here state explicitly what Formal Logic assumes to be contained or implied in the evidence afforded by any proposition, such as 'All S is P.' If we remember that in chap. iv. § 7, it was assumed that every term may have a contradictory; and if we bear in mind the principles of Contradiction and Excluded Middle, it will appear that such a proposition as 'All S is P' tells us something not only about the relations of 'S' and 'P,' but also of their relations to 'not-S' and 'not-P'; as, for example, that 'S is not not-P,' and that 'not-P is not-S.' It will be shown in the next chapter how Logicians have developed these implications in series of Immediate Inferences.

If it be asked whether it is true that every term, itself significant, has a significant contradictory, and not merely a formal contradictory, generated by force of the word 'not,' it is difficult to give any better answer than was indicated in §§ 3-5, without venturing further into Metaphysics. I shall merely say, therefore, that, granting that some such term as 'Universe' or 'Being' may have no significant contradictory, if it stand for 'whatever can be perceived or thought of'; yet every term that stands for less than 'Universe' or 'Being' has, of course, a contradictory which denotes the rest of the universe. And since every argument or train of thought is carried on within a special 'universe of discourse,' or under a certain *suppositio*, we may say that *within the given suppositio every term has a contradictory*, and that every predication concerning a term implies some predication concerning its contradictory. But the name of the *suppositio* itself has no contradictory, except with reference to a wider and inclusive *suppositio*.

The difficulty of actual reasoning, not with symbols, but about matters of fact, does not arise from the principles of Logic, but sometimes from the obscurity or complexity of the facts, sometimes from the ambiguity or clumsiness of language, sometimes from the deficiency of our own minds in penetration, tenacity and lucidity. One must do one's best to study the facts, and not be too easily discouraged.

CHAPTER VII

IMMEDIATE INFERENCES

§ 1. Under the general title of Immediate Inference Logicians discuss three subjects, namely, Opposition, Conversion, and Obversion; to which some writers add other forms, such as Whole and Part in Connotation, Contraposition, Inversion, *etc.* Of Opposition, again, all recognise four modes: Subalternation, Contradiction, Contrariety and Sub-contrariety. The only peculiarities of the exposition upon which we are now entering are, that it follows the lead of the three Laws of Thought, taking first those modes of Immediate Inference in which Identity is most important, then those which plainly involve Contradiction and Excluded Middle; and that this method results in separating the modes of Opposition, connecting Subalternation with Conversion, and the other modes with Obversion. To make up for this departure from usage, the four modes of Opposition will be brought together again in § 9.

§ 2. Subalternation.—Opposition being the relation of propositions that have the same matter and differ only in form (as A., E., I., O.), propositions of the forms A. and I. are said to be Subalterns in relation to one another, and so are E. and O.; the universal of each quality being distinguished as 'subalternans,' and the particular as 'subalternate.'

It follows from the principle of Identity that, the matter of the propositions being the same, if A. is true I. is true, and that if E. is true O. is true; for A. and E. predicate something of *All S* or *All men*; and since I. and O. make the same predication of *Some S* or *Some men*, the sense of these particular propositions has already been predicated in A. or E. If *All S is P, Some S is P*; if *No S is P, Some S is not P*; or, if *All men are fond of laughing, Some men are*; if *No men are exempt from ridicule, Some men are not*.

Similarly, if I. is false A. is false; if O. is false E. is false. If we deny any predication about *Some S*, we must deny it of *All S*; since in denying it of *Some*, we have denied it of at least part of *All*; and whatever is false in one form of words is false in any other.

On the other hand, if I. is true, we do not know that A. is; nor if O. is true, that E. is; for to infer from *Some* to *All* would be going beyond the evidence. We shall see in discussing Induction that the great problem of that part of Logic is, to determine the conditions under which we may in reality transcend this rule and infer from *Some* to *All*; though even there it will appear that, formally, the rule is observed. For the present it is enough that I. is an immediate inference from A., and O. from E.; but that A. is not an immediate inference from I., nor E. from O.

§ 3. Connotative Subalternation.—We have seen (chap. iv. § 6) that if the connotation of one term is only part of another's its denotation is greater and includes that other's. Hence genus and species stand in subaltern relation, and whatever is true of the genus is true of the species: If *All animal life is dependent on vegetation, All human life is dependent on vegetation*. On the other hand, whatever is not true of the species or narrower term, cannot be true of the whole genus: If it is false that '*All human life is happy*,' it is false that '*All animal life is happy*.'

Similar inferences may be drawn from the subaltern relation of predicates; affirming the species we affirm the genus. To take Mill's example, if *Socrates is a man, Socrates is a living creature*. On the other hand, denying the genus we deny the species: if *Socrates is not vicious, Socrates is not drunken*.

Such cases as these are recognised by Mill and Bain as immediate inferences under the principle of Identity. But some Logicians might treat them as imperfect syllogisms, requiring another premise to legitimate the conclusion, thus:

All animal life is dependent on vegetation; All human life is animal life; ∴ All human life is dependent on vegetation.

Or again:

All men are living creatures; Socrates is a man; ∴ Socrates is a living creature.

The decision of this issue turns upon the question (*cf.* chap. vi. § 3) how far a Logician is entitled to assume that the terms he uses are understood, and that the identities involved in their meanings will be recognised. And to this question, for the sake of consistency, one of two answers is required; failing which, there remains the rule of thumb. First, it may be held that no terms are understood except those that are defined in expounding the science, such as 'genus' and 'species,' 'connotation' and 'denotation.' But very few Logicians observe this limitation; few would hesitate to substitute 'not wise' for 'foolish.' Yet by what right? Malvolio being foolish, to prove that he is not-wise, we may construct the following syllogism:

Foolish is not-wise; Malvolio is foolish; ∴ Malvolio is not-wise.
Is this necessary? Why not?

Secondly, it may be held that all terms may be assumed as understood unless a definition is challenged. This principle will justify the substitution of 'not-wise' for 'foolish'; but it will also legitimate the above cases (concerning 'human life' and 'Socrates') as immediate inferences, with innumerable others that might be based upon the doctrine of relative terms: for example, *The hunter missed his aim*: therefore, *The prey escaped*. And from this principle it will further follow that all apparent syllogisms, having one premise a verbal proposition, are immediate inferences (*cf*. chap. ix. § 4).

Closely connected with such cases as the above are those mentioned by Archbishop Thomson as "Immediate Inferences by added Determinants" (*Laws of Thought*, § 87). He takes the case: '*A negro is a fellow-creature*: therefore, *A negro in suffering is a fellow-creature in suffering*.' This rests upon the principle that to increase the connotations of two terms by the same attribute or determinant does not affect the relationship of their denotations, since it must equally diminish (if at all) the denotations of both classes, by excluding the same individuals, if any want the given attribute. But this principle is true only when the added attribute is not merely the same verbally, but has the same significance in qualifying both terms. We cannot argue *A mouse is an animal*; therefore, *A large mouse is a large animal*; for 'large' is an attribute relative to the normal magnitude of the thing described.

§ 4. Conversion is Immediate Inference by transposing the terms of a given proposition without altering its quality. If the quantity is also unaltered, the inference is called 'Simple Conversion'; but if the quantity is changed from universal to particular, it is called 'Conversion by limitation' or '*per accidens*.' The given proposition is called the 'convertend'; that which is derived from it, the 'converse.'

Departing from the usual order of exposition, I have taken up Conversion next to Subalternation, because it is generally thought to rest upon the principle of Identity, and because it seems to be a good method to exhaust the forms that come only under Identity before going on to those that involve Contradiction and Excluded Middle. Some, indeed, dispute the claims of Conversion to illustrate the principle of Identity; and if the sufficient statement of that principle be 'A is A,' it may be a question how Conversion or any other mode of inference can be referred to it. But if we state it as above (chap. vi. § 3), that whatever is true in one form of words is true in any other, there is no difficulty in applying it to Conversion.

Thus, to take the simple conversion of I.,
Some S is P; ∴ Some P is S. Some poets are business-like; ∴ Some business-like men are poets.
Here the convertend and the converse say the same thing, and this is true if that is.

We have, then, two cases of simple conversion: of I. (as above) and of E. For E.:
No S is P; ∴ No P is S. No ruminants are carnivores; ∴ No carnivores are ruminants.

In converting I., the predicate (P) when taken as the new subject, being preindesignate, is treated as particular; and in converting E., the predicate (P), when taken as the new subject, is treated as universal, according to the rule in chap. v. § 1.

A. is the one case of conversion by limitation:
All S is P; ∴ Some P is S. All cats are grey in the dark; ∴ Some things grey in the dark are cats.

The predicate is treated as particular, when taking it for the new subject, according to the rule not to go beyond the evidence. To infer that *All things grey in the dark are cats* would be palpably absurd; yet no error of reasoning is commoner than the simple conversion of A. The validity of conversion by limitation may be shown thus: if, *All S is P*, then, by subalternation, *Some S is P*, and therefore, by simple conversion, *Some P is S*.

O. cannot be truly converted. If we take the proposition: *Some S is not P*, to convert this into *No P is S*, or *Some P is not S*, would break the rule in chap. vi. § 6; since *S*, undistributed in the convertend, would be distributed in the converse. If we are told that *Some men are not cooks*, we cannot infer that *Some cooks are not men*. This would be to assume that '*Some men*' are identical with '*All men*.'

By quantifying the predicate, indeed, we may convert O. simply, thus:
Some men are not cooks ∴ No cooks are some men.

And the same plan has some advantage in converting A.; for by the usual method *per accidens*, the converse of A. being I., if we convert this again it is still I., and therefore means less than our original convertend. Thus:
All S is P ∴ Some P is S ∴ Some S is P.

Such knowledge, as that *All S* (the whole of it) *is P*, is too precious a thing to be squandered in pure Logic; and it may be preserved by quantifying the predicate; for if we convert A. to Y., thus—
All S is P ∴ Some P is all S—
we may reconvert Y. to A. without any loss of meaning. It is the chief use of quantifying the predicate that, thereby, every proposition is capable of simple conversion.

The conversion of propositions in which the relation of terms is inadequately expressed (see chap. ii., § 2) by the ordinary copula (*is* or *is not*) needs a special rule. To argue thus—
A is followed by B ∴ Something followed by B is A—
would be clumsy formalism. We usually say, and we ought to say—
A is followed by B ∴ B follows A (or *is preceded by A*).

Now, any relation between two terms may be viewed from either side—*A: B* or *B: A*. It is in both cases the same fact; but, with the altered point of view, it may present a different character. For example, in the Immediate Inference—$A > B \therefore B < A$—a diminishing ratio turns into an increasing ratio, whilst the fact predicated remains the same. Given, then, a relation between two terms as viewed from one to the other, the same relation viewed from the other to the one may be called the Reciprocal. In the cases of Equality, Co-existence and Simultaneity, the given relation and its reciprocal are not only the same fact, but they also have the same character: in the cases of Greater and Less and Sequence, the character alters.

We may, then, state the following rule for the conversion of propositions in which the whole relation explicitly stated is taken as the copula: Transpose the terms, and for the given relation substitute its reciprocal. Thus—
A is the cause of B ∴ B is the effect of A.

The rule assumes that the reciprocal of a given relation is definitely known; and so far as this is true it may be extended to more concrete relations—

A is a genus of B ∴ B is a species of A   A is the father of B ∴ B is a child of A.

But not every relational expression has only one definite reciprocal. If we are told that *A is the brother of B*, we can only infer that *B is either the brother or the sister of A*. A list of all reciprocal relations is a desideratum of Logic.

§ 5. Obversion (otherwise called Permutation or Æquipollence) is Immediate Inference by changing the quality of the given proposition and substituting for its predicate the contradictory term. The given proposition is called the 'obvertend,' and the inference from it the 'obverse.' Thus the obvertend being—*Some philosophers are consistent reasoners*, the obverse will be—*Some philosophers are not inconsistent reasoners*.

The legitimacy of this mode of reasoning follows, in the case of affirmative propositions, from the principle of Contradiction, that if any term be affirmed of a subject, the contradictory term may be denied (chap. vi. § 3). To obvert affirmative propositions, then, the rule is—Insert the negative sign, and for the predicate substitute its contradictory term.

A *All S is P ∴ No S is not-P*

All men are fallible ∴ No men are infallible.

I *Some S is P ∴ some S is not-P*

Some philosophers are consistent ∴ Some philosophers are not inconsistent.

In agreement with this mode of inference, we have the rule of modern English grammar, that 'two negatives make an affirmative.'

Again, by the principle of Excluded Middle, if any term be denied of a subject, its contradictory may be affirmed: to obvert negative propositions, then, the rule is—Remove the negative sign, and for the predicate substitute its contradictory term.

E *No S is P ∴ All S is not-P*

No matter is destructible ∴ All matter is indestructible.

O *Some S is not P ∴ Some S is not-P*

Some ideals are not attainable ∴ Some ideals are unattainable.

Thus, by obversion, each of the four propositions retains its quantity but changes its quality: A. to E., I. to O., E. to A., O. to I. And all the obverses are infinite propositions, the affirmative infinites having the sense of negatives, and the negative infinites having the sense of affirmatives.

Again, having obtained the obverse of a given proposition, it may be desirable to recover the obvertend; or it may at any time be requisite to change a given infinite proposition into the corresponding direct affirmative or negative; and in such cases the process is still obversion. Thus, if *No S is not-P* be given us to recover the obvertend or to find the corresponding affirmative; the proposition being formally negative, we apply the rule for obverting negatives: 'Remove the negative sign, and for the predicate substitute its contradictory.' This yields the affirmative *All S is P*. Similarly, to obtain the obvertend of *All S is not-P*, apply the rule for obverting Affirmatives; and this yields *No S is P*.

§ 6. Contrariety.—We have seen in chap. iv. § 8, that contrary terms are such that no two of them are predicable in the same way of the same subject, whilst perhaps neither may be predicable of it. Similarly, Contrary Propositions may be defined as those of which no two are ever both true together, whilst perhaps neither may be true; or, in other words, both may be false. This is the relation between A. and E. when concerned with the same matter: as A.—*All men are wise*; E.—*No men are wise*. Such propositions cannot both be true; but they may both be false, for some men may be wise and some not. They cannot both be true; for, by the principle of Contradiction, if *wise* may be affirmed of *All men, not-wise* must be denied; but *All men are not-wise* is the obverse of *No men are wise*, which therefore may also be denied.

At the same time we cannot apply to A. and E. the principle of Excluded Middle, so as to show that one of them must be true of the same matter. For if we deny that *All men are wise*, we do not necessarily deny the attribute 'wise' of each and every man: to say that *Not all are wise* may mean no more than that *Some are not*. This gives a proposition in the form of O.; which, as we have seen, does not imply its subalternans, E.

If, however, two Singular Propositions, having the same matter, but differing in quality, are to be treated as universals, and therefore as A. and E., they are, nevertheless, contradictory and not merely contrary; for one of them must be false and the other true.

§ 7. Contradiction is a relation between two propositions analogous to that between contradictory terms (one of which being affirmed of a subject the other is denied)—such, namely, that one of them is false and the other true. This is the case with the forms A. and O., and E. and I., in the same matter. If it be true that *All men are wise*, it is false that *Some men are not wise* (equivalent by obversion to *Some men are not-wise*); or else, since the 'Some men' are included in the 'All men,' we should be predicating of the same men that they are both 'wise' and 'not-wise'; which would violate the principle of Contradiction. Similarly, *No men are wise*, being by obversion equivalent to *All men are not-wise*, is incompatible with *Some men are wise*, by the same principle of Contradiction.

But, again, if it be false that *All men are wise*, it is always true that *Some are not wise*; for though in denying that 'wise' is a predicate of 'All men' we do not deny it of each and every man, yet we deny it of 'Some men.' Of 'Some men,' therefore, by the principle of Excluded Middle, 'not-wise' is to be affirmed; and *Some men are not-wise*, is by obversion equivalent to *Some men are not wise*. Similarly, if it be false that *No men are wise*, which by obversion is equivalent to *All men are not-wise*, then it is true at least that *Some men are wise*.

By extending and enforcing the doctrine of relative terms, certain other inferences are implied in the contrary and contradictory relations of propositions. We have seen in chap. iv. that the contradictory of a given term includes all its contraries: 'not-blue,' for example, includes red and yellow. Hence, since *The sky is blue* becomes by obversion, *The sky is not not-blue*, we may also infer *The sky is not red, etc*. From the truth, then, of any proposition predicating a given term, we may infer the falsity of all propositions predicating the contrary terms in the same relation. But, on the other hand, from the falsity of a proposition predicating a given term, we cannot infer the truth of the predication of any particular contrary term. If it be false that *The sky is red*, we cannot formally infer, that *The sky is blue* (*cf.* chap. iv. § 8).

§ 8. Sub-contrariety is the relation of two propositions, concerning the same matter that may both be true but are never both false. This is the case with I. and O. If it be true that *Some men are wise*, it may also be true that *Some (other) men are not wise*. This follows from the maxim in chap. vi. § 6, not to go beyond the evidence.

For if it be true that *Some men are wise*, it may indeed be true that *All are* (this being the subalternans): and if *All are*, it is (by contradiction) false that *Some are not*; but as we are only told that *Some men are*, it is illicit to infer the falsity of *Some are not*, which could only be justified by evidence concerning *All men*.

But if it be false that *Some men are wise*, it is true that *Some men are not wise*; for, by contradiction, if *Some men are wise* is false, *No men are wise* is true; and, therefore, by subalternation, *Some men are not wise* is true.

§ 9. The Square of Opposition.—By their relations of Subalternation, Contrariety, Contradiction, and Sub-contrariety, the forms A. I. E. O. (having the same matter) are said to stand in Opposition: and Logicians represent these relations by a square having A. I. E. O. at its corners:

```
A.       Contraries        E.
  \                       /
   \  Contradictories   /
    \                 /
     \              /
      \           /
       \        /
Subalterns     Subalterns
       /        \
      /          \
     /            \
    /              \
   /  Contradictories \
  /                    \
I.      Sub-contraries    O.
```

As an aid to the memory, this diagram is useful; but as an attempt to represent the logical relations of propositions, it is misleading. For, standing at corners of the same square, A. and E., A. and I., E. and O., and I. and O., seem to be couples bearing the same relation to one another; whereas we have seen that their relations are entirely different. The following traditional summary of their relations in respect of truth and falsity is much more to the purpose:

(1) If A. is true,   I. is true,   E. is false,   O. is false.

(2) If A. is false,   I. is unknown,   E. is unknown,   O. is true.

(3) If I. is true,   A. is unknown,   E. is false,   O. is unknown.

(4) If I. is false,   A. is false,   E. is true,   O. is true.

(5) If E. is true,   A. is false,   I. is false,   O. is true.

(6) If E. is false,   A. is unknown,   I. is true,   O. is unknown.

(7) If O. is true,   A. is false,   I. is unknown,   E. is unknown.

(8) If O. is false,   A. is true,   I. is true,   E. is false.

Where, however, as in cases 2, 3, 6, 7, alleging either the falsity of universals or the truth of particulars, it follows that two of the three Opposites are unknown, we may conclude further that one of them must be true and the other false, because the two unknown are always Contradictories.

§ 10. Secondary modes of Immediate Inference are obtained by applying the process of Conversion or Obversion to the results already obtained by the other process. The best known secondary form of Immediate Inference is the Contrapositive, and this is the converse of the obverse of a given proposition. Thus:

DATUM.	OBVERSE.	CONTRAPOSITIVE.
A. All S is P	∴ No S is not-P	∴ No not-P is S
I. Some S is P	∴ Some S is not not-P	∴ (none)
E. No S is P	∴ All S is not-P	∴ Some not-P is S
O. Some S is not P	∴ Some S is not-P	∴ Some not-P is S

There is no contrapositive of I., because the obverse of I. is in the form of O., and we have seen that O. cannot be converted. O., however, has a contrapositive (*Some not-P is S*); and this is sometimes given instead of the converse, and called the 'converse by negation.'

Contraposition needs no justification by the Laws of Thought, as it is nothing but a compounding of conversion with obversion, both of which processes have already been justified. I give a table opposite of the other ways of compounding these primary modes of Immediate Inference.

		A.	I.	E.	O.
	1	All A is B.	Some A is B.	No A is B.	Some A is not B.
Obverse.	2	No A is b.	Some A is not b.	All A is b.	Some A is b.
Converse.	3	Some B is A.	Some B is A.	No B is A.	—
Obverse of Converse.	4	Some B is not a.	Some B is not a.	All B is a.	—
Contrapositive.	5	No b is A.	—	Some b is A.	Some b is A.
Obverse of Contrapositive.	6	All b is a.	—	Some b is not a.	Some b is not a.
Converse of Obverse of Converse.	7	—	—	Some a is B.	—
Obverse of Converse of Obverse of Converse.	8	—	—	Some a is not b.	—
Converse of Obverse of Contrapositive.	9	Some a is b.	—	—	—
Obverse of Converse of Obverse of Contrapositive.	10	Some a is not B.	—	—	—

In this table *a* and *b* stand for *not-A* and *not-B* and had better be read thus: for *No A is b*, *No A is not-B*; for *All b is a* (col. 6), *All not-B is not-A*; and so on.

It may not, at first, be obvious why the process of alternately obverting and converting any proposition should ever come to an end; though it will, no doubt, be considered a very fortunate circumstance that it always does end. On examining the results, it will be found that the cause of its ending is the inconvertibility of O. For E., when obverted, becomes A.; every A, when converted, degenerates into I.; every I., when obverted, becomes O.; O cannot be converted, and to obvert it again is merely to restore the former proposition: so that the whole process moves on to inevitable dissolution. I. and O. are exhausted by three transformations, whilst A. and E. will each endure seven.

Except Obversion, Conversion and Contraposition, it has not been usual to bestow special names on these processes or their results. But the form in columns 7 and 10 (*Some a is B*—*Some a is not B*), where the original predicate is affirmed or denied of the contradictory of the original subject, has been thought by Dr. Keynes to deserve a distinctive title, and he has called it the 'Inverse.' Whilst the Inverse is one form, however, Inversion is not one process, but is obtained by different processes from E. and A. respectively. In this it differs from Obversion, Conversion, and Contraposition, each of which stands for one process.

The Inverse form has been objected to on the ground that the inference *All A is B* ∴ *Some not-A is not B*, distributes *B* (as predicate of a negative proposition), though it was given as undistributed (as predicate of an affirmative proposition). But Dr. Keynes defends it on the ground that (1) it is obtained by obversions and conversions which are all legitimate and (2) that although *All A is B* does not distribute *B* in relation to *A*, it does distribute *B* in relation to some *not-A* (namely, in relation to whatever *not-A* is *not-B*). This is one reason why, in stating the rule in chap. vi. § 6, I have written: "an immediate inference ought to contain nothing that is not contained, *or formally implied*, in the proposition from which it is inferred"; and have maintained that every term formally implies its contradictory within the *suppositio*.

§ 11. Immediate Inferences from Conditionals are those which consist—(1) in changing a Disjunctive into a Hypothetical, or a Hypothetical into a Disjunctive, or either into a Categorical; and (2) in the relations of Opposition and the equivalences of Obversion, Conversion, and secondary or compound processes, which we have already examined in respect of Categoricals. As no new principles are involved, it may suffice to exhibit some of the results.

We have already seen (chap. v. § 4) how Disjunctives may be read as Hypotheticals and Hypotheticals as Categoricals. And, as to Opposition, if we recognise four forms of Hypothetical A. I. E. O., these plainly stand to one another in a Square of Opposition, just as Categoricals do. Thus A. and E. (*If A is B, C is D*, and *If A is B, C is not D*) are contraries, but not contradictories; since both may be false (*C* may sometimes be *D*, and sometimes not), though they cannot both be true. And if they are both false, their subalternates are both true, being respectively the contradictories of the universals of opposite quality, namely, I. of E., and O. of A. But in the case of Disjunctives, we cannot set out a satisfactory Square of Opposition; because, as we saw (chap. v. § 4), the forms required for E. and O. are not true Disjunctives, but Exponibles.

The Obverse, Converse, and Contrapositive, of Hypotheticals (admitting the distinction of quality) may be exhibited thus:

Datum.	Obverse.
A. If A is B, C is D	If A is B, C is not d
I. Sometimes when A is B, C is D	Sometimes when A is B, C is not d
E. If A is B, C is not D	If A is B, C is d
O. Sometimes when A is B, C is not D	Sometimes when A is B, C is d

Converse.	Contrapositive.
Sometimes when C is D, A is B	If C is d, A is not B
Sometimes when C is D, A is B	(none)
If C is D, A is not B	Sometimes when C is d, A is B
(none)	Sometimes when C is d, A is B

As to Disjunctives, the attempt to put them through these different forms immediately destroys their disjunctive character. Still, given any proposition in the form *A is either B or C*, we can state the propositions that give the sense of obversion, conversion, *etc.*, thus:

Datum.—*A is either B or C*; Obverse.—*A is not both b and c*; Converse.—*Something, either B or C, is A*; Contrapositive.—*Nothing that is both b and c is A*.

For a Disjunctive in I., of course, there is no Contrapositive. Given a Disjunctive in the form *Either A is B or C is D*, we may write for its Obverse—*In no case is A b, and C at the same time d*. But no Converse or Contrapositive of such a Disjunctive can be obtained, except by first casting it into the hypothetical or categorical form.

The reader who wishes to pursue this subject further, will find it elaborately treated in Dr. Keynes' *Formal Logic*, Part II.; to which work the above chapter is indebted.

CHAPTER VIII

ORDER OF TERMS, EULER'S DIAGRAMS, LOGICAL EQUATIONS, EXISTENTIAL IMPORT OF PROPOSITIONS

§ 1. Of the terms of a proposition which is the Subject and which the Predicate? In most of the exemplary propositions cited by Logicians it will be found that the subject is a substantive and the predicate an adjective, as in *Men are mortal*. This is the relation of Substance and Attribute which we saw (chap. i. § 5) to be the central type of relations of coinherence; and on this model other predications may be formed in which the subject is not a substance, but is treated as if it were, and could therefore be the ground of attributes; as *Fame is treacherous, The weather is changeable*. But, in literature, sentences in which the adjective comes first are not uncommon, as *Loud was the applause, Dark is the fate of man, Blessed are the peacemakers*, and so on. Here, then, 'loud,' 'dark' and 'blessed' occupy the place of the logical subject. Are they really the subject, or must we alter the order of such sentences into *The applause was loud, etc.*? If we do, and then proceed to convert, we get *Loud was the applause*, or (more scrupulously) *Some loud noise was the applause*. The last form, it is true, gives the subject a substantive word, but 'applause' has become the predicate; and if the substantive 'noise' was not implied in the first form, *Loud is the applause*, by what right is it now inserted? The recognition of Conversion, in fact, requires us to admit that, formally, in a logical proposition, the term preceding the copula is subject and the one following is predicate. And, of course, materially considered, the mere order of terms in a proposition can make no difference in the method of proving it, nor in the inferences that can be drawn from it.

Still, if the question is, how we may best cast a literary sentence into logical form, good grounds for a definite answer may perhaps be found. We must not try to stand upon the naturalness of expression, for *Dark is the fate of man* is quite as natural as *Man is mortal*. When the purpose is not merely to state a fact, but also to express our feelings about it, to place the grammatical predicate first may be perfectly natural and most effective. But the grounds of a logical order of statement must be found in its adaptation to the purposes of proof and inference. Now general propositions are those from which most inferences can be drawn, which, therefore, it is most important to establish, if true; and they are also the easiest to disprove, if false; since a single negative instance suffices to establish the contradictory. It follows that, in re-casting a literary or colloquial sentence for logical purposes, we should try to obtain a form in which the subject is distributed—

is either a singular term or a general term predesignate as 'All' or 'No.' Seeing, then, that most adjectives connote a single attribute, whilst most substantives connote more than one attribute; and that therefore the denotation of adjectives is usually wider than that of substantives; in any proposition, one term of which is an adjective and the other a substantive, if either can be distributed in relation to the other, it is nearly sure to be the substantive; so that to take the substantive term for subject is our best chance of obtaining an universal proposition. These considerations seem to justify the practice of Logicians in selecting their examples.

For similar reasons, if both terms of a proposition are substantive, the one with the lesser denotation is (at least in affirmative propositions) the more suitable subject, as *Cats are carnivores*. And if one term is abstract, that is the more suitable subject; for, as we have seen, an abstract term may be interpreted by a corresponding concrete one distributed, as *Kindness is infectious*; that is, *All kind actions suggest imitation*.

If, however, a controvertist has no other object in view than to refute some general proposition laid down by an opponent, a particular proposition is all that he need disentangle from any statement that serves his purpose.

§ 2. Toward understanding clearly the relations of the terms of a proposition, it is often found useful to employ diagrams; and the diagrams most in use are the circles of Euler.

These circles represent the denotation of the terms. Suppose the proposition to be *All hollow-horned animals ruminate*: then, if we could collect all ruminants upon a prairie, and enclose them with a circular palisade; and segregate from amongst them all the hollow-horned beasts, and enclose them with another ring-fence inside the other; one way of interpreting the proposition (namely, in denotation) would be figured to us thus:

Fig. 1.

An Universal Affirmative may also state a relation between two terms whose denotation is co-extensive. A definition always does this, as *Man is a rational animal*; and this, of course, we cannot represent by two distinct circles, but at best by one with a thick circumference, to suggest that two coincide, thus:

Fig. 2.

The Particular Affirmative Proposition may be represented in several ways. In the first place, bearing in mind that 'Some' means 'some at least, it may be all,' an I. proposition may be represented by Figs. 1 and 2; for it is true that *Some horned animals ruminate*, and that *Some men are rational*. Secondly, there is the case in which the 'Some things' of which a predication is made are, in fact, not all; whilst the predicate, though not given as distributed, yet might be so given if we wished to state the whole truth; as if we say *Some men are Chinese*. This case is also represented by Fig. 1, the outside circle representing 'Men,' and the inside one 'Chinese.' Thirdly, the predicate may appertain to some only of the subject, but to a great many other things, as in *Some horned beasts are domestic*; for it is true that some are not, and that certain other kinds of animals are, domestic. This case, therefore, must be illustrated by overlapping circles, thus:

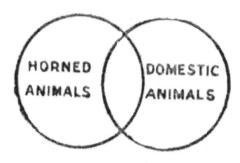

Fig. 3.

The Universal Negative is sufficiently represented by a single Fig. (4): two circles mutually exclusive, thus:

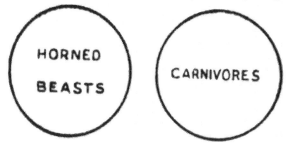

Fig. 4.

That is, *No horned beasts are carnivorous.*

Lastly, the Particular Negative may be represented by any of the Figs. 1, 3, and 4; for it is true that *Some ruminants are not hollow-horned*, that *Some horned animals are not domestic*, and that *Some horned beasts are not carnivorous.*

Besides their use in illustrating the denotative force of propositions, these circles may be employed to verify the results of Obversion, Conversion, and the secondary modes of Immediate Inference. Thus the Obverse of A. is clear enough on glancing at Figs. 1 and 2; for if we agree that whatever term's denotation is represented by a given circle, the denotation of the contradictory term shall be represented by the space outside that circle; then if it is true that *All hollow horned animals are ruminants*, it is at the same time true that *No hollow-horned animals are not-ruminants*; since none of the hollow-horned are found outside the palisade that encloses the ruminants. The Obverse of I., E. or O. may be verified in a similar manner.

As to the Converse, a Definition is of course susceptible of Simple Conversion, and this is shown by Fig. 2: 'Men are rational animals' and 'Rational animals are men.' But any other A. proposition is presumably convertible only by limitation, and this is shown by Fig. 1; where *All hollow-horned animals are ruminants*, but we can only say that *Some ruminants are hollow-horned.*

That I. may be simply converted may be seen in Fig. 3, which represents the least that an I. proposition can mean; and that E. may be simply converted is manifest in Fig. 4.

As for O., we know that it cannot be converted, and this is made plain enough by glancing at Fig. 1; for that represents the O., *Some ruminants are not hollow-horned*, but also shows this to be compatible with *All hollow-horned animals are ruminants* (A.). Now in conversion there is (by definition) no change of quality. The Converse, then, of *Some ruminants are not hollow-horned* must be a negative proposition, having 'hollow-horned' for its subject, either in E. or O.; but these would be respectively the contrary and contradictory of *All hollow-horned animals are ruminants*; and, therefore, if this be true, they must both be false.

But (referring still to Fig. 1) the legitimacy of contrapositing O. is equally clear; for if *Some ruminants are not hollow-horned*, *Some animals that are not hollow-horned are ruminants*, namely, all the animals between the two ring-fences. Similar inferences may be illustrated from Figs. 3 and 4. And the Contraposition of A. may be verified by Figs. 1 and 2, and the Contraposition of E. by Fig. 4.

Lastly, the Inverse of A. is plain from Fig. 1—*Some things that are not hollow-horned are not ruminants*, namely, things that lie outside the outer circle and are neither 'ruminants' nor 'hollow-horned.' And the Inverse of E may be studied in Fig. 4—*Some things that are not-horned beasts are carnivorous.*

Notwithstanding the facility and clearness of the demonstrations thus obtained, it may be said that a diagrammatic method, representing denotations, is not properly logical. Fundamentally, the relation asserted (or denied) to exist between the terms of a proposition, is a relation between the terms as determined by their attributes or connotation; whether we take Mill's view, that a proposition asserts that the connotation of the subject is a mark of the connotation of the predicate; or Dr. Venn's view, that things denoted by the subject (as having its connotation) have (or have not) the attribute connoted by the predicate; or, the Conceptualist view, that a judgment is a relation of concepts (that is, of connotations). With a few exceptions artificially framed (such as 'kings now reigning in Europe'), the denotation of a term is never directly and exhaustively known, but consists merely in 'all things that have the connotation.' If the value of logical training depends very much upon our habituating ourselves to construe propositions, and to realise the force of inferences from them, according to the connotation of their terms, we shall do well not to turn too hastily to the circles, but rather to regard them as means of verifying in denotation the conclusions that we have already learnt to recognise as necessary in connotation.

§ 3. The equational treatment of propositions is closely connected with the diagrammatic. Hamilton thought it a great merit of his plan of quantifying the predicate, that thereby every proposition is reduced to its true form—an equation. According to this doctrine, the proposition *All X is all Y* (U.) equates X and Y; the proposition *All X is some Y* (A.) equates X with some part of Y; and similarly with the other affirmatives (Y. and I.). And so far it is easy to follow his meaning: the Xs are identical with some or all the Ys. But, coming to the negatives, the equational interpretation is certainly less obvious. The proposition *No X is Y* (E.) cannot be said in any sense to equate X and Y; though, if we obvert it into *All X is some not-Y*, we have (in the same sense, of course, as in the above affirmative forms) X equated with part at least of 'not-Y.'

But what is that sense? Clearly not the same as that in which mathematical terms are equated, namely, in respect of some mode of quantity. For if we may say *Some X is some Y*, these Xs that are also Ys are not merely the same in number, or mass, or figure; they are the same in every respect, both quantitative and qualitative, have the same positions in time and place, are in fact identical. The proposition 2+2=4 means that any two things added to any other two are, *in respect of number*, equal to any three things added to one other thing; and this is true of all things that can be counted, however much they may differ in other ways. But *All X is all Y* means that Xs and Ys are the same things, although they have different names when viewed in different aspects or relations. Thus all equilateral triangles are equiangular triangles; but in one case they are named from the equality of their angles, and in the other from the equality of their sides. Similarly, 'British subjects' and 'subjects of King George V' are the same people, named in one case from the person of the Crown, and in the other from the Imperial Government. These logical equations, then, are in truth identities of denotation; and they are fully illustrated by the relations of circles described in the previous section.

When we are told that logical propositions are to be considered as equations, we naturally expect to be shown some interesting developments of method in analogy with the equations of Mathematics; but from Hamilton's innovations no such thing results. This cannot be said, however, of the equations of Symbolic Logic; which are the starting-point of very remarkable processes of ratiocination. As the subject of Symbolic Logic, as a whole, lies beyond the compass of this work, it will be enough to give Dr. Venn's equations corresponding with the four propositional forms of common Logic.

According to this system, universal propositions are to be regarded as not necessarily implying the existence of their terms; and therefore, instead of giving them a positive form, they are translated into symbols that express what they deny. For example, the proposition *All devils are ugly* need not imply that any such things as 'devils' really exist; but it certainly does imply that *Devils that are not ugly do not exist*. Similarly, the proposition *No angels are ugly* implies that *Angels that are ugly do not exist*. Therefore, writing x for 'devils,' y for 'ugly,' and $\bar{y}$ for 'not-ugly,' we may express A., the universal affirmative, thus:

A. $x\bar{y} = 0$.

That is, x *that is not y is nothing*; or, *Devils that are not-ugly do not exist*. And, similarly, writing x for 'angels' and y for 'ugly,' we may express E., the universal negative, thus:

E. $xy = 0$.

That is, x *that is y is nothing*; or, *Angels that are ugly do not exist*.

On the other hand, particular propositions are regarded as implying the existence of their terms, and the corresponding equations are so framed as to express existence. With this end in view, the symbol v is adopted to represent 'something,' or indeterminate reality, or more than nothing. Then, taking any particular affirmative, such as *Some metaphysicians are obscure*, and writing x for 'metaphysicians,' and y for 'obscure,' we may express it thus:

I. $xy = v$.

That is, x *that is y is something*; or, *Metaphysicians that are obscure do occur in experience* (however few they may be, or whether they all be obscure). And, similarly, taking any particular negative, such as *Some giants are not cruel*, and writing x for 'giants' and y for 'not-cruel,' we may express it thus:

O. $x\bar{y} = v$.

That is, x *that is not y is something*; or, *giants that are not-cruel do occur*—in romances, if nowhere else.

Clearly, these equations are, like Hamilton's, concerned with denotation. A. and E. affirm that the compound terms $x\bar{y}$ and xy have no denotation; and I. and O. declare that $x\bar{y}$ and xy have denotation, or stand for something. Here, however, the resemblance to Hamilton's system ceases; for the Symbolic Logic, by operating upon more than two terms simultaneously, by adopting the algebraic signs of operations, +,-, ×, ÷ (with a special signification), and manipulating the symbols by quasi-algebraic processes, obtains results which the common Logic reaches (if at all) with much greater difficulty. If, indeed, the value of logical systems were to be judged of by the results obtainable, formal deductive Logic would probably be superseded. And, as a mental discipline, there is much to be said in favour of the symbolic method. But, as an introduction to philosophy, the common Logic must hold its ground. (Venn: *Symbolic Logic*, c. 7.)

§ 4. Does Formal Logic involve any general assumption as to the real existence of the terms of propositions?

In the first place, Logic treats primarily of the *relations* implied in propositions. This follows from its being the science of proof for all sorts of (qualitative) propositions; since all sorts of propositions have nothing in common except the relations they express.

But, secondly, relations without terms of some sort are not to be thought of; and, hence, even the most formal illustrations of logical doctrines comprise such terms as S and P, X and Y, or x and y, in a symbolic or representative character. Terms, therefore, of some sort are assumed to exist (together with their negatives or contradictories) *for the purposes of logical manipulation*.

Thirdly, however, that Formal Logic cannot as such directly involve the existence of any particular concrete terms, such as 'man' or 'mountain,' used by way of illustration, is implied in the word 'formal,' that is, 'confined to what is common or abstract'; since the only thing common to all terms is to be related in some way to other terms. The actual existence of any concrete thing can only be known by experience, as with 'man' or 'mountain'; or by methodically justifiable inference from experience, as with 'atom' or 'ether.' If 'man' or 'mountain,' or 'Cuzco' be used to illustrate logical forms, they bring with them an existential import derived from experience; but this is the import of language, not of the logical forms. 'Centaur' and 'El Dorado' signify to us the non-existent; but they serve as well as 'man' and 'London' to illustrate Formal Logic.

Nevertheless, fourthly, the existence or non-existence of particular terms may come to be implied: namely, wherever the very fact of existence, or of some condition of existence, is an hypothesis or datum. Thus, given the proposition *All S is P*, to be P is made a condition of the existence of S: whence it follows that an S that is not P does not exist ($x\bar{y} = 0$). On the further hypothesis that S exists, it follows that P exists. On the hypothesis that S does not exist, the existence of P is problematic; but, then, if P does exist we cannot convert the proposition; since *Some P is S* (P existing) would involve the existence of S; which is contrary to the hypothesis.

Assuming that Universals *do not*, whilst Particulars *do*, imply the existence of their subjects, we cannot infer the subalternate (I. or O.) from the subalterns (A. or E.), for that is to ground the actual on the problematic; and for the same reason we cannot convert A. *per accidens*.

Assuming, again, a certain *suppositio* or universe, to which in a given discussion every argument shall refer, then, any propositions whose terms lie outside that *suppositio* are irrelevant, and for the purposes of that discussion are sometimes called "false"; though it seems better to call them irrelevant or meaningless, seeing that to call them false implies that they might in the same case be true. Thus propositions which, according to the doctrine of Opposition, appear to be Contradictories, may then cease to be so; for of Contradictories one is true and the other false; but, in the case supposed, both are meaningless. If the subject of discussion be Zoology, all propositions about centaurs or unicorns are absurd; and such specious Contradictories as *No centaurs play the lyre—Some centaurs do play the lyre*; or *All unicorns fight with lions—Some unicorns do not fight with lions*, are both meaningless, because in Zoology there are no centaurs nor unicorns; and, therefore, in this reference, the propositions are not really contradictory. But if the subject of discussion or *suppositio* be Mythology or Heraldry, such propositions as the above are to the purpose, and form legitimate pairs of Contradictories.

In Formal Logic, in short, we may make at discretion any assumption whatever as to the existence, or as to any condition of the existence of any particular term or terms; and then certain implications and conclusions follow in consistency with that hypothesis or datum. Still, our conclusions will themselves be only hypothetical, depending on the truth of the datum; and, of course, until this is empirically ascertained, we are as far as ever from empirical reality. (Venn: *Symbolic Logic*, c. 6; Keynes: *Formal Logic*, Part II. c. 7: *cf.* Wolf: *Studies in Logic*.)

CHAPTER IX

FORMAL CONDITIONS OF MEDIATE INFERENCE

§ 1. A Mediate Inference is a proposition that depends for proof upon two or more other propositions, so connected together by one or more terms (which the evidentiary propositions, or each pair of them, have in common) as to justify a certain conclusion, namely, the proposition in question. The type or (more properly) the unit of all such modes of proof, when of a strictly logical kind, is the Syllogism, to which we shall see that all other modes are reducible. It may be exhibited symbolically thus:

M is P; S is M: ∴ S is P.

Syllogisms may be classified, as to quantity, into Universal or Particular, according to the quantity of the conclusion; as to quality, into Affirmative or Negative, according to the quality of the conclusion; and, as to relation, into Categorical, Hypothetical and Disjunctive, according as all their propositions are categorical, or one (at least) of their evidentiary propositions is a hypothetical or a disjunctive.

To begin with Categorical Syllogisms, of which the following is an example:

All authors are vain; Cicero is an author: ∴ Cicero is vain.

Here we may suppose that there are no direct means of knowing that Cicero is vain; but we happen to know that all authors are vain and that he is an author; and these two propositions, put together, unmistakably imply that he is vain. In other words, we do not at first know any relation between 'Cicero' and 'vanity'; but we know that these two terms are severally related to a third term, 'author,' hence called a Middle Term; and thus we perceive, by mediate evidence, that they are related to one another. This sort of proof bears an obvious resemblance (though the relations involved are not the same) to the mathematical proof of equality between two quantities, that cannot be directly compared, by showing the equality of each of them to some third quantity: A = B = C ∴ A = C. Here B is a middle term.

We have to inquire, then, what conditions must be satisfied in order that a Syllogism may be formally conclusive or valid. A specious Syllogism that is not really valid is called a Parasyllogism.

§ 2. General Canons of the Syllogism.

(1) A Syllogism contains three, and no more, distinct propositions.

(2) A Syllogism contains three, and no more, distinct univocal terms.

These two Canons imply one another. Three propositions with less than three terms can only be connected in some of the modes of Immediate Inference. Three propositions with more than three terms do not show that connection of two terms by means of a third, which is requisite for proving a Mediate Inference. If we write—

All authors are vain; Cicero is a statesman—

there are four terms and no middle term, and therefore there is no proof. Or if we write—

All authors are vain; Cicero is an author: ∴ Cicero is a statesman—

here the term 'statesman' occurs without any voucher; it appears in the inference but not in the evidence, and therefore violates the maxim of all formal proof, 'not to go beyond the evidence.' It is true that if any one argued—

All authors are vain; Cicero wrote on philosophy: ∴ Cicero is vain—

this could not be called a bad argument or a material fallacy; but it would be a needless departure from the form of expression in which the connection between the evidence and the inference is most easily seen.

Still, a mere adherence to the same form of words in the expression of terms is not enough: we must also attend to their meaning. For if the same word be used ambiguously (as 'author' now for 'father' and anon for 'man of letters'), it becomes as to its meaning two terms; so that we have four in all. Then, if the ambiguous term be the Middle, no connection is shown between the other two; if either of the others be ambiguous, something seems to be inferred which has never been really given in evidence.

The above two Canons are, indeed, involved in the definition of a categorical syllogism, which may be thus stated: A Categorical Syllogism is a form of proof or reasoning (way of giving reasons) in which one categorical proposition is established by comparing two others that contain together only three terms, or that have one and only one term in common.

The proposition established, derived, or inferred, is called the Conclusion: the evidentiary propositions by which it is proved are called the Premises.

The term common to the premises, by means of which the other terms are compared, is called the Middle Term; the subject of the conclusion is called the Minor Term; the predicate of the conclusion, the Major Term.

The premise in which the minor term occurs is called the Minor Premise; that in which the major term occurs is called the Major Premise. And a Syllogism is usually written thus:

Major Premise—All authors (Middle) are vain (Major); Minor Premise—Cicero (Minor) is an author (Middle): Conclusion—∴ Cicero (Minor) is vain (Major).

Here we have three propositions with three terms, each term occurring twice. The minor and major terms are so called, because, when the conclusion is an universal affirmative (which only occurs in Barbara; see <u>chap. x. § 6</u>), its subject and predicate are respectively the less and the greater in extent or denotation; and the premises are called after the peculiar terms they contain: the expressions 'major premise' and 'minor premise' have nothing to do with the order in which the premises are presented; though it is usual to place the major premise first.

(3) No term must be distributed in the conclusion unless it is distributed in the premises.

It is usual to give this as one of the General Canons of the Syllogism; but we have seen (<u>chap. vi. § 6</u>) that it is of wider application. Indeed, 'not to go beyond the evidence' belongs to the definition of formal proof. A breech of this rule in a syllogism is the fallacy of Illicit Process of the Minor, or of the Major, according to which term has been unwarrantably distributed. The following parasyllogism illicitly distributes both terms of the conclusion:

All poets are pathetic; Some orators are not poets: ∴ No orators are pathetic.

(4) The Middle Term must be distributed at least once in the premises (in order to prove a conclusion in the given terms).

For the use of mediate evidence is to show the relation of terms that cannot be directly compared; this is only possible if the middle term furnishes the ground of comparison; and this (in Logic) requires that the whole denotation of the middle should be either included or excluded by one of the other terms; since if we only know that the other terms are related to *some* of the middle, their respective relations may not be with the same part of it.

It is true that in what has been called the "numerically definite syllogism," an inference may be drawn, though our canon seems to be violated. Thus:

60 sheep in 100 are horned; 60 sheep in 100 are blackfaced: ∴ at least 20 blackfaced sheep in 100 are horned.

But such an argument, though it may be correct Arithmetic, is not Logic at all; and when such numerical evidence is obtainable the comparatively indefinite arguments of Logic are needless. Another apparent exception is the following:

Most men are 5 feet high; Most men are semi-rational: ∴ Some semi-rational things are 5 feet high.

Here the Middle Term (men) is distributed in neither premise, yet the indisputable conclusion is a logical proposition. The premises, however, are really arithmetical; for 'most' means 'more than half,' or more than 50 per cent.

Still, another apparent exception is entirely logical. Suppose we are given, the premises—*All P is M*, and *All S is M*—the middle term is undistributed. But take the obverse of the contrapositive of both premises:

All m is p; All m is s: ∴ Some s is p.

Here we have a conclusion legitimately obtained; but it is not in the terms originally given.

For Mediate Inference depending on truly logical premises, then, it is necessary that one premise should distribute the middle term; and the reason of this may be illustrated even by the above supposed numerical exceptions. For in them the premises are such that, though neither of the two premises by itself distributes the Middle, yet they always overlap upon it. If each premise dealt with exactly half the Middle, thus barely distributing it between them, there would be no logical proposition inferrible. We require that the middle term, as used in one premise, should necessarily overlap the same term as used in the other, so as to furnish common ground for comparing the other terms. Hence I have defined the middle term as 'that term common to both premises by means of which the other terms are compared.'

(5) One at least of the premises must be affirmative; or, from two negative premises nothing can be inferred (in the given terms).

The fourth Canon required that the middle term should be given distributed, or in its whole extent, at least once, in order to afford sure ground of comparison for the others. But that such comparison may be effected, something more is requisite; the relation of the other terms to the Middle must be of a certain character. One at least of them must be, as to its extent or denotation, partially or wholly identified with the Middle; so that to that extent it may be known to bear to the other term, whatever relation we are told that so much of the Middle bears to that other term. Now, identity of denotation can only be predicated in an affirmative proposition: one premise, then, must be affirmative.

If both premises are negative, we only know that both the other terms are partly or wholly excluded from the Middle, or are not identical with it in denotation: where they lie, then, in relation to one another we have no means of knowing. Similarly, in the mediate comparison of quantities, if we are told that A and C are both of them unequal to B, we can infer nothing as to the relation of C to A. Hence the premises—

No electors are sober; No electors are independent—

however suggestive, do not formally justify us in inferring any connection between sobriety and independence. Formally to draw a conclusion, we must have affirmative grounds, such as in this case we may obtain by obverting both premises:

All electors are not-sober; All electors are not-independent: ∴ Some who are not-independent are not-sober.

But this conclusion is not in the given terms.

(6) (*a*) If one premise be negative, the conclusion must be negative: and (*b*) to prove a negative conclusion, one premise must be negative.

(*a*) For we have seen that one premise must be affirmative, and that thus one term must be partly (at least) identified with the Middle. If, then, the other premise, being negative, predicates the exclusion of the remaining term from the Middle, this remaining term must be excluded from the first term, so far as we know the first to be identical with the Middle: and this exclusion will be expressed by a negative conclusion. The analogy of the mediate comparison of quantities may here again be noticed: if A is equal to B, and B is unequal to C, A is unequal to C.

(*b*) If both premises be affirmative, the relations to the Middle of both the other terms are more or less inclusive, and therefore furnish no ground for an exclusive inference. This also follows from the function of the middle term.

For the more convenient application of these canons to the testing of syllogisms, it is usual to derive from them three Corollaries:

(i) Two particular premises yield no conclusion.

For if both premises be affirmative, *all* their terms are undistributed, the subjects by predesignation, the predicates by position; and therefore the middle term must be undistributed, and there can be no conclusion.

If one premise be negative, its predicate is distributed by position: the other terms remaining undistributed. But, by Canon 6, the conclusion (if any be possible) must be negative; and therefore its predicate, the major term, will be distributed. In the premises, therefore, both the middle and the major terms should be distributed, which is impossible: *e.g.*,

Some M is not P; Some S is M: ∴ Some S is not P.

Here, indeed, the major term is legitimately distributed (though the negative premise might have been the minor); but M, the middle term, is distributed in neither premise, and therefore there can be no conclusion.

Still, an exception may be made by admitting a bi-designate conclusion:

Some P is M; Some S is not M: ∴ Some S is not some P.

(ii) If one premise be particular, so is the conclusion.

For, again, if both premises be affirmative, they only distribute one term, the subject of the universal premise, and this must be the middle term. The minor term, therefore, is undistributed, and the conclusion must be particular.

If one premise be negative, the two premises together can distribute only two terms, the subject of the universal and the predicate of the negative (which may be the same premise). One of these terms must be the middle; the other (since the conclusion is negative) must be the major. The minor term, therefore, is undistributed, and the conclusion must be particular.

(iii) From a particular major and a negative minor premise nothing can be inferred.

For the minor premise being negative, the major premise must be affirmative (5th Canon); and therefore, being particular, distributes the major term neither in its subject nor in its predicate. But since the conclusion must be negative (6th Canon), a distributed major term is demanded, *e.g.*,

Some M is P; No S is M: ∴ ———

Here the minor and the middle terms are both distributed, but not the major (P); and, therefore, a negative conclusion is impossible.

§ 3. First Principle or Axiom of the Syllogism.—Hitherto in this chapter we have been analysing the conditions of valid mediate inference. We have seen that a single step of such inference, a Syllogism, contains, when fully expressed in language, three propositions and three terms, and that these terms must stand to one another in the relations required by the fourth, fifth, and sixth Canons. We now come to a principle which conveniently sums up these conditions; it is called the *Dictum de omni et nullo*, and may be stated thus:

Whatever is predicated (affirmatively or negatively) of a term distributed,

With which term another term can be (partly or wholly) identified,

May be predicated in like manner (affirmatively or negatively) of the latter term (or part of it).

Thus stated (nearly as by Whately in the introduction to his *Logic*) the *Dictum* follows line by line the course of a Syllogism in the First Figure (see chap. x. § 2). To return to our former example: *All authors are vain* is the same as—Vanity is predicated of all authors; *Cicero is an author* is the same as—Cicero is identified as an author; therefore *Cicero is vain*, or—Vanity may be predicated of Cicero. The *Dictum* then requires: (1) three propositions; (2) three terms; (3) that the middle term be distributed; (4) that one premise be affirmative, since only by an affirmative proposition can one term be identified with another; (5) that if one premise be negative the conclusion shall be so too, since whatever is predicated of the middle term is predicated *in like manner* of the minor.

Thus far, then, the *Dictum* is wholly analytic or verbal, expressing no more than is implied in the definitions of 'Syllogism' and 'Middle Term'; since (as we have seen) all the General Canons (except the third, which is a still more general condition of formal proof) are derivable from those definitions. However, the *Dictum* makes a further statement of a synthetic or real character, namely, that *when these conditions are fulfilled an inference is justified*; that then the major and minor terms are brought into comparison through the middle, and that the major term may be predicated affirmatively or negatively of all or part of the minor. It is this real assertion that justifies us in calling the *Dictum* an Axiom.

§ 4. Whether the Laws of Thought may not fully explain the Syllogism without the need of any synthetic principle has, however, been made a question. Take such a syllogism as the following:

All domestic animals are useful; All pugs are domestic animals: ∴ All pugs are useful.

Here (an ingenious man might urge), having once identified pugs with domestic animals, that they are useful follows from the Law of Identity. If we attend to the meaning, and remember that what is true in one form of words is true in any other form, then, all domestic animals being useful, of course pugs are. It is merely a case of subalternation: we may put it in this way:

All domestic animals are useful: ∴ Some domestic animals (*e.g.*, pugs) are useful.

The derivation of negative syllogisms from the Law of Contradiction (he might add) may be shown in a similar manner.

But the force of this ingenious argument depends on the participial clause—'having once identified pugs with domestic animals.' If this is a distinct step of the reasoning, the above syllogism cannot be reduced to one step, cannot be exhibited as mere subalternation, nor be brought directly under the law of Identity. If 'pug,' 'domestic,' and 'useful' are distinct terms; and if 'pug' and 'useful' are only known to be connected because of their relations to 'domestic': this is something more than the Laws of Thought provide for: it is not Immediate Inference, but Mediate; and to justify it, scientific method requires that its conditions be generalised. The *Dictum*, then, as we have seen, does generalise these conditions, and declares that when such conditions are satisfied a Mediate Inference is valid.

But, after all (to go back a little), consider again that proposition *All pugs are domestic animals*: is it a distinct step of the reasoning; that is to say, is it a Real Proposition? If, indeed, 'domestic' is no part of the definition of 'pug,' the proposition is real, and is a distinct part of the argument. But take such a case as this:

All dogs are useful; All pugs are dogs.

Here we clearly have, in the minor premise, only a verbal proposition; to be a dog is certainly part of the definition of 'pug.' But, if so, the inference 'All pugs are useful' involves no real mediation, and the argument is no more than this:

All dogs are useful; ∴ Some dogs (*e.g.*, pugs) are useful.
Similarly, if the major premise be verbal, thus:
All men are rational; Socrates is a man—
to conclude that 'Socrates is rational' is no Mediate Inference; for so much was implied in the minor premise, 'Socrates is a man,' and the major premise adds nothing to this.

Hence we may conclude (as anticipated in chap. vii. § 3) that 'any apparent syllogism, having one premise a verbal proposition, is really an Immediate Inference'; but that, if both premises are real propositions, the Inference is Mediate, and demands for its explanation something more than the Laws of Thought.

The fact is that to prove the minor to be a case of the middle term may be an exceedingly difficult operation (chap. xiii. § 7). The difficulty is disguised by ordinary examples, used for the sake of convenience.

§ 5. Other kinds of Mediate Inference exist, yielding valid conclusions, without being truly syllogistic. Such are mathematical inferences of Equality, as—
$A = B = C$ ∴ $A = C$.
Here, according to the usual logical analysis, there are strictly four terms—(1) A, (2) equal to B, (3) B, (4) equal to C.
Similarly with the argument *a fortiori*,
$A > B > C$ ∴ (much more) $A > C$.
This also is said to contain four terms: (1) A, (2) greater than B, (3) B, (4) greater than C. Such inferences are nevertheless intuitively sound, may be verified by trial (within the limits of sense-perception), and are generalised in appropriate axioms of their own, corresponding to the *Dictum* of the syllogism; as 'Things equal to the same thing are equal to one another,' *etc.*

Now, surely, this is an erroneous application of the usual logical analysis of propositions. Both Logic and Mathematics treat of the *relations* of terms; but whilst Mathematics employs the sign = for only one kind of relation, and for that relation exclusive of the terms; Logic employs the same signs (*is* or *is not*) for all relations, recognising only a difference of quality in predication, and treating every other difference of relation as belonging to one of the terms related. Thus Logicians read A—*is*—*equal to B*: as if *equal to B* could possibly be a term co-relative with A. Whence it follows that the argument $A = B = C$ ∴ $A = C$ contains four terms; though everybody sees that there are only three.

In fact (as observed in chap. ii. § 2) the sign of logical relation (*is* or *is not*), whilst usually adequate for class-reasoning (coinherence) and sometimes extensible to causation (because a cause implies a class of events), should never be stretched to include other relations in such a way as to sacrifice intelligence to formalism. And, besides mathematical or quantitative relations, there are others (usually considered qualitative because indefinite) which cannot be justly expressed by the logical copula. We ought to read propositions expressing time-relations (and inferences drawn accordingly) thus:

B—is before—C; A—is before—B: ∴ A—is before—C.
And in like manner *A—is simultaneous with—B; etc.* Such arguments (as well as the mathematical) are intuitively sound and verifiable, and might be generalised in axioms if it were worth while: but it is not, because no method could be founded on such axioms.

The customary use of relative terms justifies some Mediate Inferences, as, *The father of a father is a grand-father.*
Some cases, however, that at first seem obvious, are really delusive unless further data be supplied. Thus *A co-exists with B, B with C;* ∴ *A with C*—is not sound unless *B* is an instantaneous event; for where B is perdurable, *A* may co-exist with it at one time and *C* at another.

Again: *A is to the left of B, B of C;* ∴ *A of C*. This may pass; but it is not a parallel argument that if *A is north of B and B west of C*, then *A is north-west of C*: for suppose that A is a mile to the north of B, and B a yard to the west of C, then A is practically north of C; at least, its westward position cannot be expressed in terms of the mariner's compass. In such a case we require to know not only the directions but the distances of A and C from B; and then the exact direction of A from C is an affair of mathematical calculation.

Qualitative reasoning concerning position is only applicable to things in one dimension of space, or in time considered as having one dimension. Under these conditions we may frame the following generalisation concerning all Mediate Inferences: Two terms definitely related to a third, and one of them positively, are related to one another as the other term is related to the third (that is, positively or negatively); provided that the relations given are of the same kind (that is, of Time, or Coinherence, or Likeness, or Equality).

Thus, to illustrate by relations of Time—
B is simultaneous with C; A is not simultaneous with B: ∴ A is not simultaneous with C.
Here the relations are of the same kind but of different logical quality, and (as in the syllogism) a negative copula in the premises leads to a negative conclusion.

An examination in detail of particular cases would show that the above generalisation concerning all Mediate Inferences is subject to too many qualifications to be called an Axiom; it stands to the real Axioms (the *Dictum, etc.*) as the notion of the Uniformity of Nature does to the definite principles of natural order (*cf.* chap. xiii. § 8).

CHAPTER X

CATEGORICAL SYLLOGISMS

§ 1. The type of logical, deductive, mediate, categorical Inference is a Syllogism directly conformable with the *Dictum*: as—
All carnivores (M) are excitable (P); Cats (S) are carnivores (M): ∴ Cats (S) are excitable (P).
In this example P is predicated of M, a term distributed; in which term, M, S is given as included; so that P may be predicated of S.
Many arguments, however, are of a type superficially different from the above: as—
No wise man (P) fears death (M); Balbus (S) fears death (M): ∴ Balbus (S) is not a wise man (P).
In this example, instead of P being predicated of M, M is predicated of P, and yet S is given as included not in P, but in M. The divergence of such a syllogism from the *Dictum* may, however, be easily shown to be superficial by writing, instead of *No wise man fears death*, the simple, converse, *No man who fears death is wise.*

Again:

Some dogs (M) are friendly to man (P); All dogs (M) are carnivores (S): ∴ Some carnivores (S) are friendly to man (P).

Here P is predicated of M undistributed; and instead of S being included in M, M is included in S: so that the divergence from the type of syllogism to which the *Dictum* directly applies is still greater than in the former case. But if we transpose the premises, taking first

All dogs (M) are carnivores (P),

then P is predicated of M distributed; and, simply converting the other premise, we get—

Some things friendly to man (S) are dogs (M):

whence it follows that—

Some things friendly to man (S) are carnivores (P);

and this is the simple converse of the original conclusion.

Once more:

No pigs (P) are philosophers (M); Some philosophers (M) are hedonists (S): ∴ Some hedonists (S) are not pigs (P).

In this case, instead of P being predicated of M distributed, M is predicated of P distributed; and instead of S (or part of it) being included in M, we are told that some M is included in S. Still there is no real difficulty. Simply convert both the premises, and we have:

No philosophers (M) are pigs (P); Some hedonists (S) are philosophers (M).

Whence the same conclusion follows; and the whole syllogism plainly conforms directly to the *Dictum*.

Such departures as these from the normal syllogistic form are said to constitute differences of Figure (see § 2); and the processes by which they are shown to be unessential differences are called Reduction (see § 6).

§ 2. Figure is determined by the position of the Middle Term in the premises; of which position there are four possible variations. The middle term may be subject of the major premise, and predicate of the minor, as in the first example above; and this position, being directly conformable to the requirements of the *Dictum*, is called the First Figure. Or the middle term may be predicate of both premises, as in the second of the above examples; and this is called the Second Figure. Or the middle term may be subject of both premises, as in the third of the above examples; and this is called the Third Figure. Or, finally, the middle term may be predicate of the major premise, and subject of the minor, as in the fourth example given above; and this is the Fourth Figure.

It may facilitate the recollection of this most important point if we schematise the figures thus:

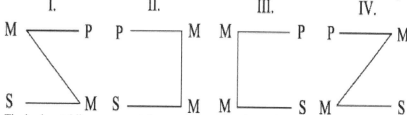

The horizontal lines represent the premises, and at the angles formed with them by the slanting or by the perpendicular lines the middle term occurs. The schema of Figure IV. resembles Z, the last letter of the alphabet: this helps one to remember it in contrast with Figure I., which is thereby also remembered. Figures II. and III. seem to stand back to back.

§ 3. The Moods of each Figure are the modifications of it which arise from different combinations of propositions according to quantity and quality. In Figure I., for example, four Moods are recognised: A.A.A., E.A.E., A.I.I., E.I.O.

A. All M is P;

A. All S is M:

A. ∴ All S is P.

E. No M is P;

A. All S is M:

E. ∴ No S is P.

A. All M is P;

I. Some S is M:

I. ∴ Some S is P.

E No M is P;

I Some S is M:

∴ Some S is not P.

Now, remembering that there are four Figures, and four kinds of propositions (A. I. E. O.), each of which propositions may be major premise, minor premise, or conclusion of a syllogism, it appears that in each Figure there may be 64 Moods, and therefore 256 in all. On examining these 256 Moods, however, we find that only 24 of them are valid (*i.e.*, of such a character that the conclusion strictly follows from the premises), whilst 5 of these 24 are needless, because their conclusions are 'weaker' or less extensive than the premises warrant; that is to say, they are particular when they might be universal. Thus, in Figure I., besides the above 4 Moods, A.A.I. and E.A.O. are valid in the sense of being conclusive; but they are superfluous, because included in A.A.A. and E.A.E. Omitting, then, these 5 needless Moods, which are called 'Subalterns' because their conclusions are subaltern (chap. vii. § 2) to those of other Moods, there remain 19 Moods that are valid and generally recognised.

§ 4. How these 19 Moods are determined must be our next inquiry. There are several ways more or less ingenious and interesting; but all depend on the application, directly or indirectly, of the Six Canons, which were shown in the last chapter to be the conditions of Mediate Inference.

(1) One way is to begin by finding what Moods of Figure I. conform to the *Dictum*. Now, the *Dictum* requires that, in the major premise, P be predicated of a term distributed, from which it follows that no Mood can be valid whose major premise is particular, as in I.A.I. or O.A.O. Again, the *Dictum* requires that the minor premise be affirmative ("with which term another is identified"); so that no Mood can be valid whose minor premise is negative, as in A.E.E. or A.O.O. By such considerations we find that in Figure I., out of 64 Moods possible, only six are valid, namely, those above-mentioned in § 3, including the two subalterns. The second step of this method is to test the Moods of the Second, Third, and Fourth Figures, by trying whether they can be reduced to one or other of the four Moods of the First (as briefly illustrated in § 1, and to be further explained in § 6).

(2) Another way is to take the above six General or Common Canons, and to deduce from them Special Canons for testing each Figure: an interesting method, which, on account of its length, will be treated of separately in the next section.

(3) Direct application of the Common Canons is, perhaps, the simplest plan. First write out the 64 Moods that are possible without regard to Figure, and then cross out those which violate any of the Canons or Corollaries, thus:

AAA, ~~AAE~~ (6th Can. *b*). AAI, ~~AAO~~ (6th Can. *b*).
~~AEA~~ (6th Can. *a*) AEE, ~~AEI~~ (6th Can. *a*) AEO,
~~AIA~~ (Cor. ii.) AIE (6th Can. *b*) AII, ~~AIO~~ (6th Can. *b*)
~~AOA~~ (6th Can. *a*) ~~AOE~~ (Cor. ii.) ~~AOI~~ (6th Can. *a*) AOO.

Whoever has the patience to go through the remaining 48 Moods will discover that of the whole 64 only 11 are valid, namely:
A.A.A., A.A.I., A.E.E., A.E.O., A.I.I., A.O.O.,
E.A.E., E.A.O., E.I.O., I.A.I., O.A.O.

These 11 Moods have next to be examined in each Figure, and if valid in every Figure there will still be 44 moods in all. We find, however, that in the First Figure, A.E.E., A.E.O., A.O.O. involve illicit process of the major term (3rd Can.); I.A.I., O.A.O. involve undistributed Middle (4th Can.); and A.A.I., E.A.O. are subalterns. In the Second Figure all the affirmative Moods, A.A.A., A.A.I., A.I.I., I.A.I., involve undistributed Middle; O.A.O. gives illicit process of the major term; and A.E.O., E.A.O. are subalterns. In the Third Figure, A.A.A., E.A.E., involve illicit process of the minor term (3rd Can.); A.E.E., A.E.O., A.O.O., illicit process of the major term. In the Fourth Figure, A.A.A. and E.A.E. involve illicit process of the minor term; A.I.I., A.O.O., undistributed Middle; O.A.O. involves illicit process of the major term; and A.E.O. is subaltern.

Those moods of each Figure which, when tried by these tests, are not rejected, are valid, namely:

Fig. I.—A.A.A., E.A.E., A.I.I., E.I.O. (A.A.I., E.A.O., Subaltern);
Fig. II.—E.A.E., A.E.E., E.I.O., A.O.O. (E.A.O., A.E.O., Subaltern);
Fig. III.—A.A.I., I.A.I., A.I.I., E.A.O., O.A.O., E.I.O.;
Fig. IV.—A.A.I., A.E.E., I.A.I., E.A.O., E.I.O. (A.E.O., Subaltern).

Thus, including subaltern Moods, there are six valid in each Figure. In Fig. III. alone there is no subaltern Mood, because in that Figure there can be no universal conclusion.

§ 5. Special Canons of the several Figures, deduced from the Common Canons, enable us to arrive at the same result by a somewhat different course. They are not, perhaps, necessary to the Science, but afford a very useful means of enabling one to thoroughly appreciate the character of formal syllogistic reasoning. Accordingly, the proof of each rule will be indicated, and its elaboration left to the reader. There is no difficulty, if one bears in mind that Figure is determined by the position of the middle term.

Fig. I., Rule (*a*): *The minor premise must be affirmative*.
For, if not, in negative Moods there will be illicit process of the major term. Applying this rule to the eleven possible Moods given in § 4, as remaining after application of the Common Canons, it eliminates A.E.E., A.E.O., A.O.O.

(*b*) *The major premise must be universal*.
For, if not, the minor premise being affirmative, the middle term will be undistributed. This rule eliminates I.A.I., O.A.O.; leaving six Moods, including two subalterns.

Fig. II. (*a*) *One premise must be negative*.
For else neither premise will distribute the middle term. This rule eliminates A.A.A., A.A.I., A.I.I., I.A.I.

(*b*) *The major premise must be universal.*

For else, the conclusion being negative, there will be illicit process of the major term. This eliminates I.A.I., O.A.O.; leaving six Moods, including two subalterns.

Fig. III. (*a*) *The minor premise must be affirmative.*

For else, in negative moods there will be illicit process of the major term. This rule eliminates A.E.E., A.E.O., A.O.O.

(*b*) *The conclusion must be particular.*

For, if not, the minor premise being affirmative, there will be illicit process of the minor term. This eliminates A.A.A., A.E.E., E.A.E.; leaving six Moods.

Fig. IV. (*a*) *When the major premise is affirmative, the minor must be universal.*

For else the middle term is undistributed. This eliminates A.I.I., A.O.O.

(*b*) *When the minor premise is affirmative the conclusion must be particular.*

Otherwise there will be illicit process of the minor term. This eliminates A.A.A., E.A.E.

(*c*) *When either premise is negative, the major must be universal.*

For else, the conclusion being negative, there will be illicit process of the major term. This eliminates O.A.O.; leaving six Moods, including one subaltern.

§ 6. Reduction is either—(1) Ostensive or (2) Indirect. Ostensive Reduction consists in showing that an argument given in one Mood can also be stated in another; the process is especially used to show that the Moods of the second, third, and fourth Figures are equivalent to one or another Mood of the first Figure. It thus proves the validity of the former Moods by showing that they also essentially conform to the *Dictum*, and that all Categorical Syllogisms are only superficial varieties of one type of proof.

To facilitate Reduction, the recognised Moods have all had names given them; which names, again, have been strung together into mnemonic verses of great force and pregnancy:

Barbara, Celarent, Darii, Ferioque prioris: Cesare, Camestres, Festino, Baroco, secundæ: Tertia, Darapti, Disamis, Datisi, Felapton, Bocardo, Ferison, habet: Quarta insuper addit Bramantip, Camenes, Dimaris, Fesapo, Fresison.

In the above verses the names of the Moods of Fig. I. begin with the first four consonants B, C, D, F, in alphabetical order; and the names of all other Moods likewise begin with these letters, thus signifying (except in Baroco and Bocardo) the mood of Fig. I., to which each is equivalent, and to which it is to be reduced: as Bramantip to Barbara, Camestres to Celarent, and so forth.

The vowels A, E, I, O, occurring in the several names, give the quantity and quality of major premise, minor premise, and conclusion in the usual order.

The consonants s and p, occurring after a vowel, show that the proposition which the vowel stands for is to be converted either (s) simply or (p) *per accidens*; except where s or p occurs after the third vowel of a name, the conclusion: then it refers not to the conclusion of the given Mood (say Disamis), but to the conclusion of that Mood of the first Figure to which the given Mood is reduced (Darii).

M (*mutare*, metathesis) means 'transpose the premises' (as of Ca*m*estres).

C means 'substitute the contradictory of the conclusion for the foregoing premise,' a process of the Indirect Reduction to be presently explained (see Baroco, § 8).

The other consonants, r, n, t (with b and d, when not initial), occurring here and there, have no mnemonic significance.

What now is the problem of Reduction? The difference of Figures depends upon the position of the Middle Term. To reduce a Mood of any other Figure to the form of the First, then, we must so manipulate its premises that the Middle Term shall be subject of the major premise and predicate of the minor premise.

Now in Fig. II. the Middle Term is predicate of both premises; so that the minor premise may need no alteration, and to convert the major premise may suffice. This is the case with Cesare, which reduces to Celarent by simply converting the major premise; and with Festino, which by the same process becomes Ferio. In Camestres, however, the minor premise is negative; and, as this is impossible in Fig. I., the premises must be transposed, and the new major premise must be simply converted: then, since the transposition of the premises will have transposed the terms of the conclusion (according to the usual reading of syllogisms), the new conclusion must be simply converted in order to prove the validity of the original conclusion. The process may be thus represented (*s.c.* meaning 'simply convert')

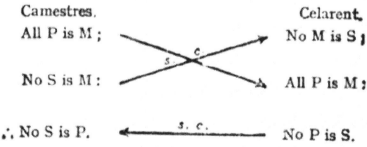

The Ostensive Reduction of Baroco also needs special explanation; for as it used to be reduced indirectly, its name gives no indication of the ostensive process. To reduce it ostensively let us call it Faksnoko, where k means 'obvert the foregoing premise.' By thus obverting (k) and simply converting (s) (in sum, contraposing) the major premise, and obverting the minor premise, we get a syllogism in Ferio, thus:

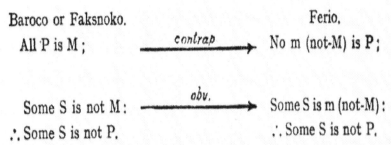

In Fig. III. the middle term is subject of both premises; so that, to reduce its Moods to the First Figure, it may be enough to convert the minor premise. This is the case with Darapti, Datisi, Felapton, and Ferison. But, with Disamis, since the major premise must in the First Figure be universal, we must transpose the premises, and then simply convert the new minor premise; and, lastly, since the major and minor terms have now changed places, we must simply convert the new conclusion in order to verify the old one. Thus:

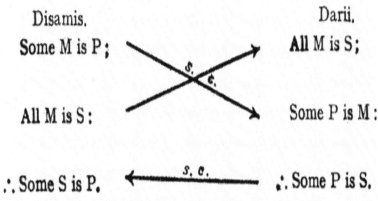

Bocardo, like Baroco, indicates by its name the indirect process. To reduce it ostensively let its name be Doksamrosk, and proceed thus:

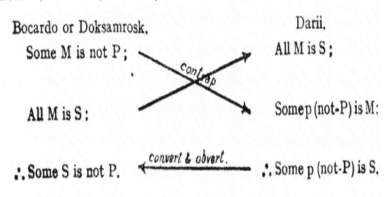

In Fig. IV. the position of the middle term is, in both premises, the reverse of what it is in the First Figure; we may therefore reduce its Moods either by transposing the premises, as with Bramantip, Camenes, and Dimaris; or by converting both premises, the course pursued with Fesapo and Fresison. It may suffice to illustrate by the case of Bramantip:

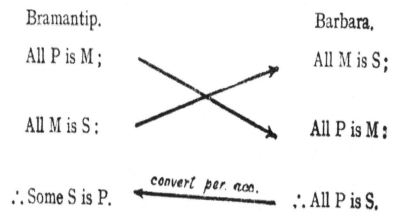

This case shows that a final significant consonant (s, p, or sk) in the name of any Mood refers to the conclusion of the new syllogism in the First Figure; since p in Bramantip cannot refer to that Mood's own conclusion in I.; which, being already particular, cannot be converted *per accidens*.

Finally, in Fig. I., Darii and Ferio differ respectively from Barbara and Celarent only in this, that their minor premises, and consequently their conclusions, are subaltern to the corresponding propositions of the universal Moods; a difference which seems insufficient to give them rank as distinct forms of demonstration. And as for Barbara and Celarent, they are easily reducible to one another by obverting their major premises and the new conclusions, thus:

Barbara.   **Celarent.**

All M is P ;   ──── obv. ────▶   No M is p (not-P) ;

All S is M :   ─────────────▶   All S is M :

∴ All S is P.   ◀──── obv ────   ∴ No S is p (not-P).

There is, then, only one fundamental syllogism.

§ 7. A new version of the mnemonic lines was suggested in *Mind* No. 27, with the object of (1) freeing them from all meaningless letters, (2) showing by the name of each Mood the Figure to which it belongs, (3) giving names to indicate the ostensive reduction of Baroco and Bocardo. To obtain the first two objects, *l* is used as the mark of Fig. I., *n* of Fig II., *r* of Fig. III., *t* of Fig. IV. The verses (to be scanned discreetly) are as follows:

Balala,	Celalel,	Dalil,	Felioque prioris:
			{Faksnoko}
Cesane,	Camenes,	Fesinon,	{Banoco,} secundæ:
Tertia,	Darapri,	Drisamis,	Darisi, Ferapro,
Doksamrosk	}, Ferisor habet:	Quarta insuper addit.	
Bocaro	}		
Bamatip,	Cametes,	Dimatis,	Fesapto, Fesistot.

De Morgan praised the old verses as "more full of meaning than any others that ever were made"; and in defence of the above alteration it may be said that they now deserve that praise still more.

§ 8. Indirect reduction is the process of proving a Mood to be valid by showing that the supposition of its invalidity involves a contradiction. Take Baroco, and (since the doubt as to its validity is concerned not with the truth of the premises, but with their relation to the conclusion) assume the premises to be true. Then, if the conclusion be false, its contradictory is true. The conclusion being in O., its contradictory will be in A. Substituting this A. for the minor premise of Baroco, we have the premises of a syllogism in Barbara, which will be found to give a conclusion in A., contradictory of the original minor premise; thus:

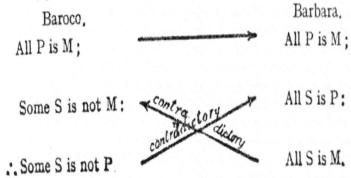

But the original minor premise, *Some S is not M*, is true by hypothesis; and therefore the conclusion of Barbara, *All S is M*, is false. This falsity cannot, however, be due to the form of Barbara, which we know to be valid; nor to the major premise, which, being taken from Baroco, is true by hypothesis: it must, therefore, lie in the minor premise of Barbara, *All S is P*; and since this is contradictory of the conclusion of Baroco *Some S is not P*, that conclusion was true.

Similarly, with Bocardo, the Indirect Reduction proceeds by substituting for the major premise the contradictory of the conclusion; thus again obtaining the premises of a syllogism in Barbara, whose conclusion is contradictory of the original major premise. Hence the initial B in Baroco and Bocardo: it points to a syllogism in Barbara as the means of Indirect Reduction (*Reductio ad impossibile*).

Any other Mood may be reduced indirectly: as, for example, Dimaris. If this is supposed to be invalid and the conclusion false, substitute the contradictory of the conclusion for the major premise, thus obtaining the premises of Celarent:

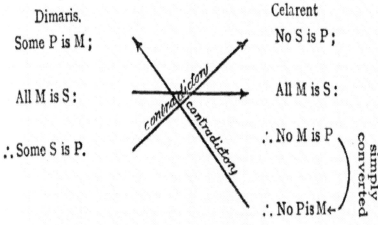

The conclusion of Celarent, simply converted, contradicts the original major premise of Dimaris, and is therefore false. Therefore the major premise of Celarent is false, and the conclusion of Dimaris is true. We might, of course, construct mnemonic names for the Indirect Reduction of all the Moods: the name of Dimaris would then be Cicari.

§ 9. The need or use of any Figure but the First has been much discussed by Logicians. Since, in actual debate, arguments are rarely stated in syllogistic form, and, therefore, if reduced to that form for closer scrutiny, generally have to be treated with some freedom; why not always throw them at once into the First Figure? That Figure has manifest advantages: it agrees directly with the *Dictum*; it gives conclusions in all four propositional forms, and therefore serves every purpose of full affirmation or denial, of showing agreement or difference (total or partial), of establishing the contradictories of universal statements; and it is the only Figure in which the subject and predicate of the conclusion occupy the same positions in the premises, so that the course of argument has in its mere expression an easy and natural flow.

Still, the Second Figure also has a very natural air in some kinds of negative arguments. The parallelism of the two premises, with the middle term as predicate in both, brings out very forcibly the necessary difference between the major and minor terms that is involved in their opposite relations to the middle term. *P is not, whilst S is, M*, says Cesare: that drives home the conviction that *S is not P*. Similarly in Camestres: *Deer do, oxen do not, shed their horns*. What is the conclusion?

The Third Figure, again, furnishes in Darapti and Felapton, the most natural forms of stating arguments in which the middle term is singular:

Socrates was truthful; Socrates was a Greek: ∴ Some Greek was truthful.

Reducing this to Fig I., we should get for the minor premise, *Some Greek was Socrates*: which is certainly inelegant. Still, it might be urged that, in relation to proof, elegance is an extraneous consideration. And as for the other advantage claimed for Fig. III.—that, as it yields only particular conclusions, it is useful in establishing contradictories against universals—for that purpose none of its Moods can be better than Darii or Ferio.

As for Fig. IV., no particular advantage has been claimed for it. It is of comparatively late recognition (sometimes called the 'Galenian,' after Galen, its supposed discoverer); and its scientific claim to exist at all is disputed. It is said to be a mere inversion of Fig. I.; which is not true in any sense in which Figs. II. and III. may not be condemned as partial inversions of Fig. I., and as having therefore still less claim to recognition. It is also said to invert the order of thought; as if thought had only one order, or as if the order of thought had anything to do with Formal Logic. Surely, if distinction of Figure be recognised at all, the Fourth Figure is scientifically necessary, because it is inevitably generated by an analysis of the possible positions of the middle term.

§ 10. Is Reduction necessary, however; or have not all the Figures equal and independent validity? In one sense not only every Figure but each Mood has independent validity: for any one capable of abstract thinking sees its validity by direct inspection; and this is true not only of the abstract Moods, but very frequently of particular concrete arguments. But science aims at unifying knowledge; and after reducing all possible arguments that form categorical syllogisms to the nineteen Moods, it is another step in the same direction to reduce these Moods to one form. This is the very nature of science: and, accordingly, the efforts of some Logicians to expound separate principles of each Figure seem to be supererogatory. Grant that they succeed; and what can the next step be, but either to reduce these principles to the *Dictum*, or the *Dictum* and the rest to one of these principles? Unless this can be done there is no science of Formal Logic. If it is done, what is gained by reducing the principles of the other Figures to the *Dictum*, instead of the Moods of the other Figures to those of the first Figure? It may, perhaps, be said that to show (1) that the Moods of the second, third, and fourth Figures flow from their own principles (though, in fact, these principles are laboriously adapted to the Moods); and (2) that these principles may be derived from the *Dictum*, is the more uncompromisingly gradual and regular method: but is not Formal Logic already sufficiently encumbered with formalities?

§ 11. Euler's diagrams are used to illustrate the syllogism, though not very satisfactorily, thus:

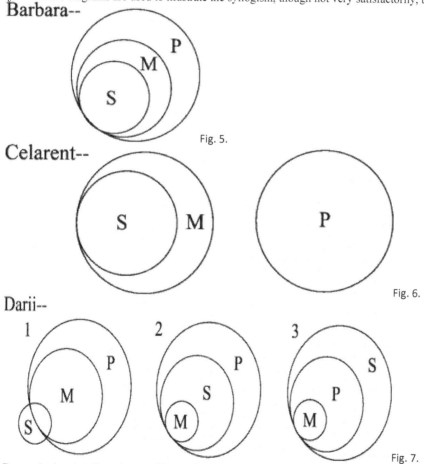

Remembering that 'Some' means 'It may be all,' it is plain that any one of these diagrams in Fig. 7, or the one given above for Barbara, may represent the denotative relations of P, M and S in Darii; though no doubt the diagram we generally think of as representing Darii is No. 1 in Fig. 7.

Remembering that A may be U, and that, therefore, wherever A occurs there may be only one circle for S and P, these syllogisms may be represented by only two circles, and Barbara by only one.

Ferio--

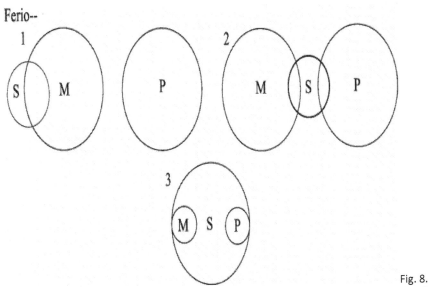

Fig. 8.

Here, again, probably, we generally think of No. 1 as the diagram representing Ferio; but 2, or 3, or that given above for Celarent, is compatible with the premises.

If instead of dealing with M, P, and S, a concrete example be taken of Darii or Ferio, a knowledge of the facts of the case will show what diagram is suitable to it. But, then, surely it must be possible to do without the diagram. These diagrams, of course, can be used to illustrate Moods of the other Figures.

CHAPTER XI
ABBREVIATED AND COMPOUND ARGUMENTS

§ 1. In ordinary discussion, whether oral or written, it is but rarely that the forms of Logic are closely adhered to. We often leave wide gaps in the structure of our arguments, trusting the intelligence of those addressed to bridge them over; or we invert the regular order of propositions, beginning with the conclusion, and mentioning the premises, perhaps, a good while after, confident that the sagacity of our audience will make all smooth. Sometimes a full style, like Macaulay's, may, by means of amplification and illustration, spread the elements of a single syllogism over several pages—a pennyworth of logic steeped in so much eloquence. These practices give a great advantage to sophists; who would find it very inconvenient to state explicitly in Mood and Figure the pretentious antilogies which they foist upon the public; and, indeed, such licences of composition often prevent honest men from detecting errors into which they themselves have unwittingly fallen, and which, with the best intentions, they strive to communicate to others: but we put up with these drawbacks to avoid the inelegance and the tedium of a long discourse in accurate syllogisms.

Many departures from the strictly logical statement of reasonings consist in the use of vague or figurative language, or in the substitution for one another of expressions supposed to be equivalent, though, in fact, dangerously discrepant. Against such occasions of error the logician can provide no safeguard, except the advice to be careful and discriminating in what you say or hear. But as to any derangement of the elements of an argument, or the omission of them, Logic effectually aids the task of restoration; for it has shown what the elements are that enter into the explicit statement of most ratiocinations, namely, the four forms of propositions and what that connected order of propositions is which most easily and surely exposes the validity or invalidity of reasoning, namely, the premises and conclusion of the Syllogism. Logic has even gone so far as to name certain abbreviated forms of proof, which may be regarded as general types of those that actually occur in debate, in leading articles, pamphlets and other persuasive or polemic writings—namely, the Enthymeme, Epicheirema and Sorites.

§ 2. The Enthymeme, according to Aristotle, is the Syllogism of probable reasoning about practical affairs and matters of opinion, in contrast with the Syllogism of theoretical demonstration upon necessary grounds. But, as now commonly treated, it is an argument with one of its elements omitted; a Categorical Syllogism, having one or other of its premises, or else its conclusion, suppressed. If the major premise be suppressed, it is called an Enthymeme of the First Order; if the minor premise be wanting, it is said to be of the Second Order; if the conclusion be left to be understood, there is an Enthymeme of the Third Order.

Let the following be a complete Syllogism:
All free nations are enterprising; The Dutch are a free nation: ∴ The Dutch are enterprising.
Reduced to Enthymemes, this argument may be put thus:
In the First Order:
The Dutch are a free nation: ∴ The Dutch are enterprising.

In the Second Order—
All free nations are enterprising; ∴ The Dutch are enterprising.
In the Third Order—
All free nations are enterprising; And the Dutch are a free nation.

It is certainly very common to meet with arguments whose statement may be represented by one or other of these three forms; indeed, the Enthymeme is the natural substitute for a full syllogism in oratory: whence the transition from Aristotle's to the modern meaning of the term. The most unschooled of men readily apprehend its force; and a student of Logic can easily supply the proposition that may be wanted in any case to complete a syllogism, and thereby test the argument's formal validity. In any Enthymeme of the Third Order, especially, to supply the conclusion cannot present any difficulty at all; and hence it is a favourite vehicle of innuendo, as in Hamilton's example:

Every liar is a coward; And Caius is a liar.

The frankness of this statement and its reticence, together, make it a biting sarcasm upon Caius.

The process of finding the missing premise in an Enthymeme of either the First or the Second Order, so as to constitute a syllogism, is sometimes called Reduction; and for this a simple rule may be given: Take that term of the given premise which does not occur in the conclusion (and which must therefore be the Middle), and combine it with that term of the conclusion which does not occur in the given premise; the proposition thus formed is the premise which was requisite to complete the Syllogism. If the premise thus constituted contain the predicate of the conclusion, the Enthymeme was of the First Order; if it contain the subject of the conclusion, the Enthymeme was of the Second Order.

That a statement in the form of a Hypothetical Proposition may really be an Enthymeme (as observed in chap. v. § 4) can easily be shown by recasting one of the above Enthymemes thus: *If all free nations are enterprising, the Dutch are enterprising*. Such statements should be treated according to their true nature.

To reduce the argument of any ordinary discourse to logical form, the first care should be to make it clear to oneself what exactly the conclusion is, and to state it adequately but as succinctly as possible. Then look for the evidence. This may be of an inductive character, consisting of instances, examples, analogies; and, if so, of course its cogency must be evaluated by the principles of Induction, which we shall presently investigate. But if the evidence be deductive, it will probably consist of an Enthymeme, or of several Enthymemes one depending on another. Each Enthymeme may be isolated and expanded into a syllogism. And we may then inquire: (1) whether the syllogisms are formally correct according to Barbara (or whatever the appropriate Mood); (2) whether the premises, or the ultimate premises, are true in fact.

§ 3. A Monosyllogism is a syllogism considered as standing alone or without relation to other arguments. But, of course, a disputant may be asking to prove the premises of any syllogism; in which case other syllogisms may be advanced for that purpose. When the conclusion of one syllogism is used to prove another, we have a chain-argument which, stated at full length, is a Polysyllogism. In any Polysyllogism, again, a syllogism whose conclusion is used as the premise of another, is called in relation to that other a Prosyllogism; whilst a syllogism one of whose premises is the conclusion of another syllogism, is in relation to that other an Episyllogism. Two modes of abbreviating a Polysyllogism, are usually discussed, the Epicheirema and the Sorites.

§ 4. An Epicheirema is a syllogism for one or both of whose premises a reason is added; as—

All men are mortal, for they are animals; Socrates is a man, for rational bipeds are men: ∴ Socrates is mortal.

The Epicheirema is called Single or Double, says Hamilton, according as an "adscititious proposition" attaches to one or both of the premises. The above example is of the double kind. The Single Epicheirema is said to be of the First Order, if the adscititious proposition attach to the major premise; if to the minor, of the Second Order. (Hamilton's *Logic*: Lecture xix.)

An Epicheirema, then, is an abbreviated chain of reasoning, or Polysyllogism, comprising an Episyllogism with one or two enthymematic Prosyllogisms. The major premise in the above case, *All men are mortal, for they are animals*, is an Enthymeme of the First Order, suppressing its own major premise, and may be restored thus:

All animals are mortal; All men are animals: ∴ All men are mortal.

The minor premise, *Socrates is a man, for rational bipeds are men*, is an Enthymeme of the Second Order, suppressing its own minor premise, and may be restored thus:

All rational bipeds are men; Socrates is a rational biped: ∴ Socrates is a man.

§ 5. The Sorites is a Polysyllogism in which the Conclusions, and even some of the Premises, are suppressed until the arguments end. If the chain of arguments were freed of its enthymematic character, the suppressed conclusions would appear as premises of Episyllogisms.

Two varieties of Sorites are recognised, the Aristotelian (so called, though not treated of by Aristotle), and the Goclenian (named after its discoverer, Goclenius of Marburg, who flourished about 1600 A.D.). In order to compare these two forms of argument, it will be convenient to place side by side Hamilton's classical examples of them.

Aristotelian.	Goclenian.
Bucephalus is a horse;	An animal is a substance;
A horse is a quadruped;	A quadruped is an animal;
A quadruped is an animal;	A horse is a quadruped;
An animal is a substance:	Bucephalus is a horse:
Bucephalus is a substance.	Bucephalus is a substance.

The reader wonders what is the difference between these two forms. In the Aristotelian Sorites the minor term occurs in the first premise, and the major term in the last; whilst in the Goclenian the major term occurs in the first premise, and the minor in the last. But since the character of premises is fixed by their terms, not by the order in which they are written, there cannot be a better example of a distinction without a difference. At a first glance, indeed, there may seem to be a more important point involved; the premises of the Aristotelian Sorites seem to proceed in the order of Fig. IV. But if that were really so the conclusion would be, *Some Substance is Bucephalus*. That, on the contrary, every one writes the conclusion, *Bucephalus is a substance*, proves that the logical order of the premises is in Fig. I. Logically, therefore, there is absolutely no difference between these two forms, and pure reason requires either that the "Aristotelian Sorites" disappear from the text-books, or that it be regarded as in Fig. IV., and its conclusion converted. It is the shining merit of Goclenius to have restored the premises of the Sorites to the usual order of Fig. I.: whereby he has raised to himself a monument more durable than brass, and secured indeed the very cheapest immortality.

The common Sorites, then, being in Fig. I., its rules follow from those of Fig. I:

(1) Only one premise can be particular; and, if any, only that in which the minor term occurs.

For, just as in Fig I., a particular premise anywhere else involves undistributed Middle.

(2) Only one premise can be negative; and, if any, only that in which the major term occurs.

For if there were two negative premises, at the point where the second entered the chain of argument there must be a syllogism with two negative premises, which is contrary to Rule 5; whilst if one premise be negative it must be that which contains the major term, for the same reason as in Fig. I., namely, that the conclusion will be negative, and that therefore only a negative major premise can prevent illicit process of the major term.

If we expand a Sorites into its constituent syllogisms, the conclusions successively suppressed will reappear as major premises; thus:

(1) An animal is a substance;

A quadruped is an animal:

∴ A quadruped is a substance.

(2) A quadruped is a substance;

A horse is a quadruped:

∴ A horse is a substance.

(3) A horse is a substance:

Bucephalus is a horse:

∴ Bucephalus is a substance.

This suffices to show that the Protosyllogism of a Goclenian Sorites is an Enthymeme of the Third Order; after which the argument is a chain of Enthymemes of the First Order, or of the First and Third combined, since the conclusions as well as the major premises are omitted, except in the last one.

Lest it should be thought that the Sorites is only good for arguments so frivolous as the above, I subjoin an example collected from various parts of Mill's *Political Economy*:—

The cost of labour depends on the efficiency of labour; The rate of profits depends on the cost of labour; The investment of capital depends on the rate of profits; Wages depend on the investment of capital: ∴ Wages depend on the efficiency of labour.

Had it occurred to Mill to construct this Sorites, he would have modified his doctrine of the wages-fund, and would have spared many critics the malignant joy of refuting him.

§ 6. The Antinomy is a combination of arguments by which contradictory attributes are proved to be predicable of the same subject. In symbols, thus:

All M is P;   All N is p;

All S is M: All S is N:

∴ All S is P.   ∴ All S is p.

Now, by the principle of Contradiction, S cannot be P and p (not-P): therefore, if both of the above syllogisms are sound, S, as the subject of contradictory attributes, is logically an impossible thing. The contradictory conclusions are called, respectively, Thesis and Antithesis.

To come to particulars, we may argue: (1) that a constitution which is at once a monarchy, an aristocracy and a democracy, must comprise the best elements of all three forms; and must, therefore, be the best of all forms of government: the British Constitution is, therefore, the best of all. But (2) such a constitution must also comprise the worst elements of monarchy, aristocracy and democracy; and, therefore, must be the worst of all forms. Are we, then, driven to conclude that the British Constitution, thus proved to be both the best and worst, does not really exist at all, being logically impossible? The proofs seem equally cogent; but perhaps neither the best nor the worst elements of the simpler constitutions need be present in our own in sufficient force to make it either good or bad.

Again:

(1) Every being who is responsible for his actions is free;

Man is responsible for his actions:

∴ Man is free.

(2) Every being whose actions enter into the course of nature is not free;

Man is such a being:

∴ Man is not free.

Does it, then, follow that 'Man,' as the subject of contradictory attributes, is a nonentity? This doctrine, or something like it, has been seriously entertained; but if to any reader it seem extravagant (as it certainly does to me), he will no doubt find an error in the above arguments. Perhaps the major term is ambiguous.

For other examples it is enough to refer to the *Critique of Pure Reason*, where Kant sets out the Antinomies of Rational Cosmology. But even if we do not agree with Kant that the human understanding, in attempting to deal with certain subjects beyond its reach, inevitably falls into such contradictory reasonings; yet it can hardly be doubted that we not unfrequently hold opinions which, if logically developed, result in Antinomies. And, accordingly, the Antinomy, if it cannot be imputed to Reason herself, may be a very fair, and a very wholesome *argumentum ad hominem*. It was the favourite weapon of the Pyrrhonists against the dogmatic philosophies that flourished after the death of Aristotle.

CHAPTER XII
CONDITIONAL SYLLOGISMS

§ 1. Conditional Syllogisms may be generally described as those that contain conditional propositions. They are usually divided into two classes, Hypothetical and Disjunctive.

A Hypothetical Syllogism is one that consists of a Hypothetical Major Premise, a Categorical Minor Premise, and a Categorical Conclusion. Two Moods are usually recognised the *Modus ponens*, in which the antecedent of the hypothetical major premise is affirmed; and the *Modus tollens*, in which its consequent is denied.

(1) *Modus ponens*, or Constructive.

If A is B, C is D; A is B: ∴ C is D.

If Aristotle's reasoning is conclusive, Plato's theory of Ideas is erroneous;

Aristotle's reasoning is conclusive: ∴ Plato's theory of Ideas is erroneous.

Rule of the *Modus ponens*: The antecedent of the major premise being affirmed in the minor premise, the consequent is also affirmed in the conclusion.

(2) *Modus tollens*, or Destructive.

If A is B, C is D; C is not D: ∴ A is not B.

If Pythagoras is to be trusted, Justice is a number; Justice is not a number: ∴ Pythagoras is not to be trusted.

Rule of the *Modus tollens*: The consequent of the major premise being denied in the minor premise, the antecedent is denied in the conclusion.

By using negative major premises two other forms are obtainable: then, either by affirming the antecedent or by denying the consequent, we draw a negative conclusion.

Thus (*Modus ponens*):

If A is B, C is not D;

A is B:

∴ C is not D.

(*Modus tollens*):

If A is B, C is not D;

C is D:

∴ A is not B.

Further, since the antecedent of the major premise, taken by itself, may be negative, it seems possible to obtain four more forms, two in each Mood, from the following major premises:

(1) If A is not B, C is D; (2) If A is not B, C is not D.

But since the quality of a Hypothetical Proposition is determined by the quality of its consequent, not at all by the quality of its antecedent, we cannot get from these two major premises any really new Moods, that is to say, Moods exhibiting any formal difference from the four previously expounded.

It is obvious that, given the hypothetical major premise—

If A is B, C is D—

we cannot, by denying the antecedent, infer a denial of the consequent. That A is B, is a mark of C being D; but we are not told that it is the sole and indispensable condition of it. If men read good books, they acquire knowledge; but they may acquire knowledge by other means, as by observation. For the same reason, we cannot by affirming the consequent infer the affirmation of the antecedent: Caius may have acquired knowledge; but we cannot thence conclude that he has read good books.

To see this in another light, let us recall chap. v. § 4, where it was shown that a hypothetical proposition may be translated into a categorical one; whence it follows that a Hypothetical Syllogism may be translated into a Categorical Syllogism. Treating the above examples thus, we find that the *Modus ponens* (with affirmative major premise) takes the form of Barbara, and the *Modus tollens* the form of Camestres:

Modus ponens.	Barbara.
If A is B, C is D;	The case of A being B is a case of C being D;
A is B:	This is a case of A being B:
∴ C is D.	∴ This is a case of C being D.

Now if, instead of this, we affirm the consequent, to form the new minor premise,

This is a case of C being D,

there will be a Syllogism in the Second Figure with two affirmative premises, and therefore the fallacy of undistributed Middle. Again:

Modus tollens.	Camestres.
If A is B, C is D;	The case of A being B is a case of C being D:
C is not D:	This is not a case of C being D:
∴ A is not B.	∴ This is not a case of A being B.

But if, instead of this, we deny the antecedent, to form the new minor premise,

This is not a case of A being B,

there arises a syllogism in the First Figure with a negative minor premise, and therefore the fallacy of illicit process of the major term.

By thus reducing the Hypothetical Syllogism to the Categorical form, what is lost in elegance is gained in intelligibility. For, first, we may justify ourselves in speaking of the hypothetical premise as the major, and of the categorical premise as the minor; since in the categorical form they contain respectively the major and minor terms. And, secondly, we may justify ourselves in treating the Hypothetical Syllogism as a kind of Mediate Inference, in spite of the fact that it does not exhibit two terms compared by means of a third; since in the Categorical form such terms distinctly appear: a new term ('This') emerges in the position of the minor; the place of the Middle is filled by the antecedent of the major premise in the *Modus ponens*, and by the consequent in the *Modus tollens*.

The mediate element of the inference in a Hypothetical Syllogism consists in asserting, or denying, the fulfilment of a given condition; just as in a Categorical syllogism to identify the minor term with the Middle is a condition of the major term's being predicated of it. In the hypothetical proposition—

If A is B, C is D—

the Antecedent, *A is B*, is the *conditio sufficiens*, or mark, of the Consequent, *C is D*; and therefore the Consequent, *C is D*, is a *conditio sine qua non* of the antecedent, *A is B*; and it is by means of affirming the former condition, or else denying the latter, that a conclusion is rendered possible.

Indeed, we need not say that the element of mediation consists in affirming, *or denying*, the fulfilment of a given condition: it is enough to say 'in affirming.' For thus to explain the *Modus tollens*, reduce it to the *Modus ponens* (contrapositing the major premise and obverting the minor):

	Celarent.
If A is B, C is D:	The case of C being not-D is not a case of A being B;
∴ If C is not-D, A is not B;	
C is not-D:	This is a case of C being not-D:
∴ A is not B.	∴ This is not a case of A being B.

The above four forms commonly treated of as Hypothetical Syllogisms, are called by Ueberweg and Dr. Keynes 'Hypothetico-Categorical.' Ueberweg restricts the name 'Hypothetical' simply (and Dr. Keynes the name 'Conditional') to such Syllogisms as the following, having two Hypothetical Premises:

If C is D, E is F; If A is B, C is D: ∴ If A is B, E is F.

If we recognise particular hypothetical propositions (see chap. v. § 4), it is obvious that such Syllogisms may be constructed in all the Moods and Figures of the Categorical Syllogism; and of course they may be translated into Categoricals. We often reason in this hypothetical way. For example:

If the margin of cultivation be extended, rents will rise; If prices of produce rise, the margin of cultivation will be extended: ∴ If prices of produce rise, rents will rise.

But the function of the Hypothetical Syllogism (commonly so called), as also of the Disjunctive Syllogism (to be discussed in the next section) is to get rid of the conditional element of the premises, to pass from suspense to certainty, and obtain a decisive categorical conclusion; whereas these Syllogisms with two hypothetical premises leave us still with a hypothetical conclusion. This circumstance seems to ally them more closely with Categorical Syllogisms than with those that are discussed in the present chapter. That they are Categoricals in disguise may be seen by considering that the above syllogism is not materially significant, unless in each proposition the word 'If' is equivalent to 'Whenever.' Accordingly, the name 'Hypothetical Syllogism,' is here employed in the older usage.

§ 2. A Disjunctive Syllogism consists of a Disjunctive Major Premise, a Categorical Minor Premise, and a Categorical Conclusion.

How many Moods are to be recognised in this kind of argument depends on whether the alternatives of the Disjunctive Premise are regarded as mutually exclusive or possibly coincident. In saying '*Either* A is B, *or* C is D,' do we mean 'either, but not both,' or 'either, it may be both'? (See chap. v. § 4.)

When the alternatives of the Disjunctive are not exclusive, we have only the

Modus tollendo ponens.

Either A is B, or C is D;

A is not B (or C is not D):

∴ C is D (or A is B).

Either wages fall, or the weaker hands are dismissed;
Wages do not fall: ∴ The weaker hands are dismissed.
But we cannot argue—
Wages fall: ∴ The weaker hands are not dismissed;
since in 'hard times' both events may happen together.
Rule of the *Modus tollendo ponens*: If one alternative be denied, the other is affirmed.
When, however, the alternatives of the Disjunctive are mutually exclusive, we have also the

Modus ponendo tollens.

Either A is B, or C is D;

A is B (or C is D):

∴ C is not D (or A is not B).

Either the Tories or the Whigs win the election;
The Tories win: ∴ The Whigs do not win.
We may also, of course, argue as above in the *Modus tollendo ponens*—
The Tories do not win: ∴ The Whigs do.

But in this example, to make the *Modus tollendo ponens* materially valid, it must be impossible that the election should result in a tie. The danger of the Disjunctive Proposition is that the alternatives may not, between them, exhaust the possible cases. Only contradictory alternatives are sure to cover the whole ground.

Rule of the *Modus ponendo tollens:* If one alternative be affirmed, the other is denied.

Since a disjunctive proposition may be turned into a hypothetical proposition (chap. v. § 4,) a Disjunctive Syllogism may be turned into a Hypothetical Syllogism:

Modus tollendo ponens.	*Modus ponens.*
Either A is B, or C is D;	If A is not B, C is D;
A is not B:	A is not B:

∴ C is D.   ∴ C is D.

Similarly the *Modus ponendo tollens* is equivalent to that kind of *Modus ponens* which may be formed with a negative major premise; for if the alternatives of a disjunctive proposition be exclusive, the corresponding hypothetical be affirmative or negative:

Modus ponendo tollens.	Modus ponens.
Either A is B, or C is D;	If A is B, C is not D;
A is B:	A is B:
∴ C is not D.	∴ C is not D.

Hence, finally, a Disjunctive Syllogism being equivalent to a Hypothetical, and a Hypothetical to a Categorical; a Disjunctive Syllogism is equivalent and reducible to a Categorical. It is a form of Mediate Inference in the same sense as the Hypothetical Syllogism is; that is to say, the conclusion depends upon an affirmation, or denial, of the fulfilment of a condition implied in the disjunctive major premise.

§ 3. The Dilemma is perhaps the most popularly interesting of all forms of proof. It is a favourite weapon of orators and wits; and "impaled upon the horns of a dilemma" is a painful situation in which every one delights to see his adversary. It seems to have been described by Rhetoricians before finding its way into works on Logic; and Logicians, to judge from their diverse ways of defining it, have found some difficulty in making up their minds as to its exact character.

There is a famous Dilemma employed by Demosthenes, from which the general nature of the argument may be gathered:

If Æschines joined in the public rejoicings, he is inconsistent; if he did not, he is unpatriotic; But either he joined, or he did not join: Therefore he is either inconsistent or unpatriotic.

That is, reduced to symbols:

If A is B, C is D; and if E is F, G is H: But either A is B, or E is F; ∴ Either C is D or G is H (*Complex Constructive*).

This is a compound Conditional Syllogism, which may be analysed as follows:

Either A is B or E is F.

Suppose that E is not F:	Suppose that A is not B:
Then A is B.	Then E is F.
But if A is B, C is D;	But if E is F, G is H;
(A is B):	(E is F):
∴ C is D.	∴ G is H.

∴ Either C is D or G is H.

A Dilemma, then, is a compound Conditional Syllogism, having for its Major Premise two Hypothetical Propositions, and for its Minor Premise a Disjunctive Proposition, whose alternative terms either affirm the Antecedents or deny the Consequents of the two Hypothetical Propositions forming the Major Premise.

The hypothetical propositions in the major premise, may have all four terms distinct (as in the above example); and then the conclusion is a disjunctive proposition, and the Dilemma is said to be Complex. Or the two hypothetical propositions may have a common antecedent or a common consequent; and then the conclusion is a categorical proposition, and the Dilemma is said to be Simple.

Again, the alternatives of the disjunctive minor premise may be affirmative or negative: if affirmative, the Dilemma is called Constructive; and if negative, Destructive.

Using, then, only affirmative hypothetical propositions in the major premise, there are four Moods:

1. The Simple Constructive—

If A is B, C is D; and if E is F, C is D: But either A is B, or E is F: ∴ C is D.

If the Tories win the election, the Government will avoid innovation; and if the Whigs win, the House of Lords will prevent them innovating: But either the Tories or the Whigs will win: ∴ There will be no innovation.

2. The Complex Constructive—

If A is B, C is D; and if E is F, G is H: But either A is B, or E is F: ∴ Either C is D or G is H.

If appearance is all that exists, reality is a delusion; and if there is a substance beyond consciousness, knowledge of reality is impossible: But either appearance is all, or there is a substance beyond consciousness: ∴ Either reality is a delusion, or a knowledge of it is impossible.

3. Simple Destructive—

If A is B, C is D; and if A is B, E is F: But either C is not D, or E is not F: ∴ A is not B.

If table-rappers are to be trusted, the departed are spirits; and they also exert mechanical energy: But either the departed are not spirits, or they do not exert mechanical energy: ∴ Table-rappers are not to be trusted.

4. Complex Destructive—

If A is B, C is D; and if E is F, G is H: But either C is not D, or G is not H: ∴ Either A is not B, or E is not F.

If poetic justice is observed, virtue is rewarded; and if the mirror is held up to Nature, the villain triumphs: But either virtue is not rewarded, or the villain does not triumph: ∴ Either poetic justice is not observed, or the mirror is not held up to Nature.

Such are the four Moods of the Dilemma that emerge if we only use affirmative hypotheticals for the major premise; but, certainly, it is often quite as natural to employ two negative hypotheticals (indeed, one might be affirmative and the other negative; but waive that); and then four more moods emerge, all having negative conclusions. It is needless to intimidate the reader by drawing up these four moods in battle array: they always admit of reduction to the foregoing moods by obverting the hypotheticals. Still, by the same process we may greatly decrease the number of moods of the Categorical Syllogism; and just as some Syllogisms are most simply expressed in Celarent or Cesare, so some Dilemmas are most simply stated with negative major premises—e.g., The example of a Simple Constructive Dilemma above given would run more naturally thus: *If the Tories win, the Government will not innovate; and if the Whigs, the Lords will not let them*: and similarly Demosthenes' Dilemma—*If Æschines joined, he is not consistent; and if he did not, he is not patriotic*. Moreover, the propriety of recognising Dilemmas with negative major premises, follows from the above analysis of the Dilemma into a combination of Conditional Syllogisms, even if (as in § 1 of this chapter) we take account of only four Moods of the Hypothetical Syllogism.

In the rhetorical use of the Dilemma, it may be observed that the disjunction in the minor premise ought to be obvious, or (at any rate) easily acceptable to the audience. Thus, *Either the Tories or the Whigs will win; Either Æschines joined in the rejoicings, or he did not*; such propositions are not likely to be disputed. But if the orator must stop to prove his minor premise, the smacking effect of this figure (if the expression be allowed) will be lost. Hence the minor premises of other examples given above are only fit for a select audience. That *Either ghosts are not spirits, or they do not exert mechanical energy*, supposes a knowledge of the principle, generally taught by physical philosophers, that only matter is the vehicle of energy; and that *Either appearance is all, or there is substance beyond consciousness*, is a doctrine which only metaphysical philosophers could be expected to understand, and upon which they could not be expected to agree. However, the chief danger is that a plausible disjunction may not be really such as to exclude any middle ground: *Either the Tories or the Whigs win*, is bad, if a tie be possible; though in the above argument this is negligible, seeing that a tie cannot directly cause innovations. *Either Æschines joined in the rejoicings, or he did not*, does not allow for a decent conformity with the public movement where resistance would be vain; yet such conformity as need not be inconsistent with subsequent condemnation of the proceedings, nor incompatible with patriotic reserve founded on a belief that the rejoicings are premature and ominous.

Another rhetorical consideration is, that the alternatives of the disjunctive conclusion of a Complex Dilemma should both point the same way, should be equally distasteful or paradoxical. 'Either inconsistent or unpatriotic': horrid words to a politician! 'Either no reality or no possible knowledge of it': very disappointing to an anxious inquirer! Thus the disjunctive conclusion is as bad for an opponent as the categorical one in a Simple Dilemma.

Logicians further speak of the Trilemma, with three Hypotheticals and a corresponding triple Disjunction; and of a Polylemma, with any further number of perplexities. But anyone who has a taste for logical forms may have it amply gratified in numerous text-books.

CHAPTER XIII

TRANSITION TO INDUCTION

§ 1. Having now discussed Terms, Propositions, Immediate and Mediate Inferences, and investigated the conditions of formal truth or consistency, we have next to consider the conditions of material truth: whether (or how far) it is possible to arrive at propositions that accurately represent the course of nature or of human life. Hitherto we have dealt with no sort of proof that gives any such assurance. A valid syllogism guarantees the truth of its conclusion, provided the premises be true: but what of the premises? The relation between the premises of a valid syllogism and its conclusion is the same as the relation between the antecedent and consequent of a hypothetical proposition. If A is B, C is D: grant that A is B, and it follows that C is D; and, similarly, grant the premises of a syllogism, and the conclusion follows. Again, grant that C is not D, and it follows that A is not B; and, similarly, if the conclusion of a valid syllogism be false, it follows that one, or other, or both of the premises must be false. But, once more, grant that C is D, and it does not follow that A is B; so neither, if the conclusion of a syllogism be true, does it follow that the premises are. For example:—

Sociology is an exact science; Mathematics is a branch of Sociology: ∴ Mathematics is an exact science.

Here the conclusion is true although the premises are absurd. Or again:—

Mathematics is an exact science; Sociology is a branch of Mathematics: ∴ Sociology is an exact science.

Here the major premise is true, but the minor is false, and the conclusion is false. In both cases, however, whether the conclusion be true or false, it equally follows from the premises, if there is any cogency in Barbara. The explanation of this is, that Barbara has only formal cogency; and that whether the conclusion of that, or any other valid mood, shall be true according to fact and experience, depends upon how the form is filled up. How to establish the premises, then, is a most important problem; and it still remains to be solved.

§ 2. We may begin by recalling the distinction between the denotation and connotation of a general term: the denotation comprising the things or events which the term is a name for; the connotation comprising the common qualities on account of which these things are called by the same name. Obviously, there are very few general terms whose denotation is exhaustively known; since the denotation of a general term comprises all the things that have its connotation, or that ever have had, or that ever will have it, whether they exist here, or in Australia, or in the Moon, or in the utmost stars. No one has examined all men, all mammoths, all crystals, all falling bodies, all cases of fever, all revolutions, all stars—nor even all planets, since from time to time new ones are discerned. We have names for animals that existed long before there were men to observe them, and of which we know only a few bones, the remains of multitudinous species; and for others that may continue to exist when men have disappeared from the earth.

If, indeed, we definitely limit the time, or place, or quantity of matter to be explored, we may sometimes learn, within the given limits, all that there is to know: as all the bones of a particular animal, or the list of English monarchs hitherto, or the names of all the members of the House of Commons at the present time. Such cases, however, do not invalidate the above logical truth that few general terms are exhaustively known in their denotation; for the very fact of assigning limits of time and place impairs the generality of a term. The bones of a certain animal may be all examined, but not the bones of all animals, nor even of one species. The English monarchs that have reigned hitherto may be known, but there may be many still to reign.

The general terms, then, with which Logic is chiefly concerned, the names of Causes and Kinds, such as gravitation, diseases, social events, minerals, plants and animals, stand for some facts that are, or have been, known, and for a great many other similar ones that have not been, and never will be, known. The use of a general term depends not upon our direct knowledge of everything comprised in its denotation, but upon our readiness to apply it to anything that has its connotation, whether we have seen the thing or not, and even though we never can perceive it; as when a man talks freely of the ichthyosaurus, or of the central heat of planets, or of atoms and ether.

Hence Universal Propositions, which consist of general terms, deceive us, if we suppose that their predicates are directly known to be related to all the facts denoted by their subjects. In exceptional cases, in which the denotation of a subject is intentionally limited, such exhaustive direct knowledge may be possible; as that "all the bones of a certain animal consist of phosphate of lime," or that every member of the present Parliament wears a silk hat. But what predication is possible concerning the hats of all members of Parliament from the beginning? Ordinarily, then, whilst the relation of predicate to subject has been observed in some cases, in much the greater number of cases our belief about it depends upon something besides observation, or may be said (in a certain sense) to be taken on trust.

'All rabbits are herbivorous': why do we believe that? We may have seen a few wild rabbits feeding: or have kept tame ones, and tried experiments with their diet; or have read of their habits in a book of Natural History; or have studied the anatomy and physiology of the digestive system in many sorts of animals: but with whatever care we add testimony and scientific method to our own observation, it still remains true that the rabbits observed by ourselves and others are few in comparison with those that live, have lived and will live. Similarly of any other universal proposition; that it 'goes beyond the evidence' of direct observation plainly follows from the fact that the general terms, of which such propositions consist, are never exhaustively known in their denotation. What right have we then to state Universal Propositions? That is the problem of Inductive Logic.

§ 3. Universal Propositions, of course, cannot always be proved by syllogisms; because to prove a universal proposition by a syllogism, its premises must be universal propositions; and, then, these must be proved by others. This process may sometimes go a little way, thus: *All men are mortal*, because *All animals are*; and *All animals are mortal*, because *All composite bodies are subject to dissolution*. Were there no limit to such sorites, proof would always involve a *regressus ad infinitum*, for which life is too short; but, in fact, prosyllogisms soon fail us.

Clearly, the form of the Syllogism must itself be misleading if the universal proposition is so: if we think that premises prove the conclusion because they themselves have been established by detailed observation, we are mistaken. The consideration of any example will show this. Suppose any one to argue:

All ruminants are herbivorous; Camels are ruminants: ∴ Camels are herbivorous.

Have we, then, examined all ruminants? If so, we must have examined all camels, and cannot need a syllogism to prove their herbivorous nature: instead of the major premise proving the conclusion, the proof of the conclusion must then be part of the proof of the major premise. But if we have not examined all ruminants, having omitted most giraffes, most deer, most oxen, etc., how do we know that the unexamined (say, some camels) are not exceptional? Camels are vicious enough to be carnivorous; and indeed it is said that Bactrian camels will eat flesh rather than starve, though of course their habit is herbivorous.

Or, again, it is sometimes urged that—

All empires decay: ∴ Britain will decay.

This is manifestly a prediction: at present Britain flourishes, and shows no signs of decay. Yet a knowledge of its decay seems necessary, to justify any one in asserting the given premise. If it is a question whether Britain will decay, to attempt (while several empires still flourish) to settle the matter by asserting that *all* empires decay, seems to be 'a begging of the question.' But although this latter case is a manifest prediction, it does not really differ from the former one; for the proof that camels are herbivorous has no limits in time. If valid, it shows not only that they are, but also that they will be, herbivorous.

Hence, to resort to a dilemma, it may be urged: If *all* the facts of the major premise of any syllogism have been examined, the syllogism is needless; and if *some* of them have not been examined, it is a *petitio principii*. But either all have been examined, or some have not. Therefore; the syllogism is either useless or fallacious.

§ 4. A way of escape from this dilemma is provided by distinguishing between the formal and material aspects of the syllogism considered as a means of proof. It begs the question formally, but not materially; that is to say, if it be a question whether camels are herbivorous, and to decide it we are told that '*all* ruminants are,' laying stress upon the 'all,' as if all had been examined, though in fact camels have not been, then the question as to camels is begged. The form of a universal proposition is then offered as evidence, when in fact the evidence has not been universally ascertained. But if in urging that 'all ruminants are herbivorous' no more is meant than that so many other ruminants of different species are known to be herbivorous, and that the ruminant stomach is so well adapted to a coarse vegetable diet, that the same habit may be expected in other ruminants, such as camels, the argument then rests upon material evidence without unfairly implying the case in question. Now the nature of the material evidence is plainly this, that the resemblance of camels to deer, oxen, *etc.*, in chewing the cud, justifies us in believing that they have a further resemblance in feeding on herbs; in other words, we assume that *resemblance is a ground of inference*.

Another way of putting this difficulty which we have just been discussing, with regard to syllogistic evidence, is to urge that by the Laws of Syllogism a conclusion must never go beyond the premises, and that therefore no progress in knowledge can ever be established, except by direct observation. Now, taking the syllogism formally, this is true: if the conclusion go beyond the premises, there must be either four terms, or illicit process of the major or minor term. But, taking it materially, the conclusion may cover facts which were not in view when the major premise was laid down; facts of which we predicate something not as the result of direct observation, but because they resemble in a certain way those facts which had been shown to carry the predicate when the major premise was formed.

'What sort of resemblance is a sufficient ground of inference?' is, therefore, the important question alike in material Deduction and in Induction; and in endeavouring to answer it we shall find that the surest ground of inference is resemblance of causation. For example, it is due to causation that ruminants are herbivorous. Their instincts make them crop the herb, and their stomachs enable them easily to digest it; and in these characters camels are like the other ruminants.

§ 5. In ch. ix, § 3, the *Dictum de omni et nullo* was stated: 'Whatever may be predicated of a term distributed may be predicated of anything that can be identified with that term.' Nothing was there said (as nothing was needed) of the relations that might be implied in the predication. But now that it comes to the ultimate validity of predication, we must be clear as to what these relations are; and it will also be convenient to speak no longer of terms, as in Formal Logic, but of the things denoted. What relations, then, can be determined between concrete facts or phenomena (physical or mental) with the greatest certainty of general truth; and what axioms are there that sanction mediate inferences concerning those relations?

In his *Logic* (B. II. c. 2, § 3) Mill gives as the axiom of syllogistic reasoning, instead of the *Dictum*: "A thing which co-exists with another thing, which other co-exists with a third thing, also co-exists with that third thing." Thus the peculiar properties of Socrates co-exist with the attributes of man, which co-exist with mortality: therefore, Socrates is mortal. But, again, he says that the ground of the syllogism is Induction; that man is mortal is an induction. And, further, the ground of Induction is causation; the law of causation is the ultimate major premise of every sound induction. Now causation is the principle of the succession of phenomena: how, then, can the syllogism rest on an axiom concerning co-existence? On reflection, too, it must appear that 'Man is mortal' predicates causation: the human constitution issues in death.

The explanation of this inconsistency may perhaps be found in the history of Mill's work. Books I. and II. were written in 1831; but being unable at that time to explain Induction, he did not write Book III. until 1837-8. Then, no doubt, he revised the earlier Books, but not enough to bring his theory of the syllogism into complete agreement with the theory of Induction; so that the axiom of co-existence was allowed to stand.

Mill also introduced the doctrine of Natural Kinds as a ground of Induction supplementary, at least provisionally, to causation; and to reasoning about Kinds, or Substance and Attribute, his axiom of co-existence is really adapted. Kinds are groups of things that agree amongst themselves and differ from all others in a multitude of qualities: these qualities co-exist, or co-inhere, with a high degree of constancy; so that where some are found others may be inferred. Their co-inherence is not to be considered an ultimate fact; for, "since everything which occurs is determined by laws of causation and collocations of the original causes, it follows that the co-existences observable amongst effects cannot themselves be the subject of any similar set of laws distinct from laws of causation" (B. III. c. 5, § 9). According to the theory of evolution (worked out since Mill wrote), Kinds—that is, species of plants, animals and minerals—with their qualities are all due to causation. Still, as we can rarely, or never, trace the causes with any fullness or precision, a great deal of our reasoning, as, *e.g.*, about men and camels, does in fact trust to the relative permanence of natural Kinds as defined by co-inhering attributes.

To see this more clearly, we should consider that causation and natural Kinds are not at present separable; propositions about causation in concrete phenomena (as distinct from abstract 'forces') always involve the assumption of Kinds. For example—'Water rusts iron,' or the oxygen of water combines with iron immersed in it to form rust: this statement of causation assumes that water, oxygen, iron, and iron-rust are known Kinds. On the other hand, the constitution of every concrete thing, and manifestly of every organised body, is always undergoing change, that is, causation, upon which fact its properties depend.

How, then, can we frame principles of mediate reasoning, about such things? So far as we consider them as Kinds, it is enough to say: *Whatever can be identified as a specimen of a known substance or Kind has the properties of that Kind.* So far as we consider them as in the relation of causation, we may say: *Whatever relation of events can be identified with the relation of cause and effect is constant.* And these principles may be generalised thus: *Whatever is constantly related to a phenomenon (cause or Kind), determined by certain characters, is related in the same way to any phenomenon, that has the same characters.* Taking this as axiom of the syllogism materially treated, we see that herbivorousness, being constantly related to ruminants, is constantly related to camels; mortality to man and, therefore, to Socrates; rusting to the immersion of iron in water generally and, therefore, to this piece of iron. *Nota notæ, nota rei ipsius* is another statement of the same principle; still another is Mill's axiom, "Whatever has a mark has what it is a mark of." A mark is anything (A) that is never found without something else (B)—a phenomenon constantly related to another phenomenon—so that wherever A is found, B may be expected: human nature is a mark of mortality.

§ 6. The Syllogism has sometimes been discarded by those who have only seen that, as formally stated, it is either useless or fallacious: but those who also perceive its material grounds retain and defend it. In fact, great advantages are gained by stating an argument as a formal syllogism. For, in the first place, we can then examine separately the three conditions on which the validity of the argument depends:

(1) Are the Premises so connected that, *if they are true*, the Conclusion follows? This depends upon the formal principles of chap. x.

(2) Is the Minor Premise true? This question can only arise when the minor premise is a real proposition; and then it may be very difficult to answer. Water rusts iron; but is the metal we are now dealing with a fair specimen of iron? Few people, comparatively, know how to determine whether diamonds, or even gold or silver coins, are genuine. That *Camels are ruminants* is now a verbal proposition to a Zoologist, but not to the rest of us; and to the Zoologist the ascertaining of the relation in which camels stand to such ruminants as oxen and deer, was not a matter of analysing words but of dissecting specimens. What a long controversy as to whether the human race constitutes a Family of the Primates! That 'the British Empire is an empire' affords no matter for doubt or inquiry; but how difficult to judge whether the British Empire resembles Assyria, Egypt, Rome, Spain in those characters and circumstances that caused their downfall!

(3) Is the Major Premise true? Are all ruminants herbivorous? If there be any exceptions to the rule, camels are likely enough to be among the exceptions. And here the need of Inductive Logic is most conspicuous: how can we prove our premises when they are universal propositions? Universal propositions, however, are also involved in proving the minor premise: to prove a thing to be iron, we must know the constant reactions of iron.

A second advantage of the syllogism is, that it makes us fully aware of what an inference implies. An inference must have some grounds, or else it is a mere prejudice; but whatever the grounds, if sufficient in a particular case, they must be sufficient for all similar cases, they must admit of being generalised; and to generalise the grounds of the inference, is nothing else than to state the major premise. If the evidence is sufficient to justify the argument that camels are herbivorous *because* they are ruminants, it must also justify the major premise, *All ruminants are herbivorous*; for else the inference cannot really depend merely upon the fact of ruminating. To state our evidence syllogistically, then, must be possible, if the evidence is mediate and of a logical kind; and to state it in this formal way, as depending on the truth of a general principle (the major premise), increases our sense of responsibility for the inference that is thus seen to imply so much; and if any negative instances lie within our knowledge, we are the more likely to remember them. The use of syllogisms therefore tends to strengthen our reasonings.

A third advantage is, that to formulate an accurate generalisation may be useful to others: it is indeed part of the systematic procedure of science. The memoranda of our major premises, or reasons for believing anything, may be referred to by others, and either confirmed or refuted. When such a memorandum is used for further inferences, these inferences are said, in the language of Formal Logic, to be drawn *from* it, as if the conclusion were contained in our knowledge of the major premise; but, considering the limited extent of the material evidence, it is better to say that the inference is drawn *according to* the memorandum or major premise, since the grounds of the major premise and of the conclusion are in fact the same (Mill: *Logic*, B. II. c. 3). Inductive proofs may be stated in Syllogisms, and inductive inferences are drawn *according to* the Law of Causation.

§ 7. To assume that resemblance is a ground of inference, and that substance and attribute, or cause and effect, are phenomena constantly related, implies belief in the Uniformity of Nature. The Uniformity of Nature cannot be defined, and is therefore liable to be misunderstood. In many ways Nature seems not to be uniform: there is great variety in the sizes, shapes, colours and all other properties of things: bodies falling in the open air—pebbles, slates, feathers—descend in different lines and at different rates; the wind and weather are proverbially uncertain; the course of trade or of politics, is full of surprises. Yet common maxims, even when absurd, testify to a popular belief that the relations of things are constant: the doctrine of St. Swithin and the rhyme beginning 'Evening red and morning grey,' show that the weather is held to be not wholly unpredictable; as to human affairs, it is said that 'a green Yule makes a fat churchyard,' that 'trade follows the flag,' and that 'history repeats itself'; and Superstition knows that witches cannot enter a stable-door if a horse-shoe is nailed over it, and that the devil cannot cross a threshold inscribed with a perfect pentagram. But the surest proof of a belief in the uniformity of nature is given by the conduct of men and animals; by that adherence to habit, custom and tradition, to which in quiet times they chiefly owe their safety, but which would daily disappoint and destroy them, if it were not generally true that things may be found where they have been left and that in similar circumstances there are similar events.

Now this general belief, seldom distinctly conceived, for the most part quite unconscious (as a principle), merely implied in what men do, is also the foundation of all the Sciences; which are entirely occupied in seeking the Laws (that is, the Uniformities) of Nature. As the uniformity of nature cannot be defined, it cannot be proved; the most convincing evidence in its favour is the steady progress made by Science whilst trusting in it. Nevertheless, what is important is not the comprehensive but indeterminate notion of Uniformity so much as a number of First Principles, which may be distinguished in it as follows:

(1) The Principles of Contradiction and Excluded Middle (ch. vi. § 3) declare that in a given relation to a given phenomenon any two or more other phenomena are incompatible (B is not A and a); whilst the given phenomenon either stands related to another phenomenon or not (B is either A or a). It is not only a matter of Logic but of fact that, if a leaf is green, it is not under the same conditions red or blue, and that if it is not green it is some other colour.

(2) Certain Axioms of Mediate Evidence: as, in Mathematics, 'that magnitudes equal to the same magnitude are equal to one another'; and, in Logic, the *Dictum* or its material equivalent.

(3) That all Times and all Spaces are commensurable; although in certain relations of space (as π) the unit of measurement must be infinitely small.—If Time really trotted with one man and galloped with another, as it seems to; if space really swelled in places, as De Quincey dreamed that it did; life could not be regulated, experience could not be compared and science would be impossible. The Mathematical Axioms would then never be applicable to space or time, or to the objects or processes that fill them.

(4) The Persistence of Matter and Energy: the physical principle that, in all changes of the universe, the quantities of Matter and Energy (actual and potential, so-called) remain the same.—For example, as to matter, although dew is found on the grass at morning without any apparent cause, and although a candle seems to burn away to a scrap of blackened wick, yet every one knows that the dew has been condensed from vapour in the air, and that the candle has only turned into gas and smoke. As to energy, although a stone thrown up to the housetop and resting there has lost actual energy, it has gained such a position that the slightest touch may bring it to the earth again in the same time as it took to travel upwards; so on the house-top it is said to have potential energy. When a boiler works an engine, every time the piston is thrust forward (mechanical energy), an equivalent in heat (molecular energy) is lost. But for the elucidation of these principles, readers must refer to treatises of Chemistry and Physics.

(5) Causation, a special form of the foregoing principles of the persistence of matter and energy, we shall discuss in the next chapter. It is not to be conceived of as anything occult or noumenal, but merely as a special mode of the uniformity of Nature or experience.

(6) Certain Uniformities of Co-existence; but for want of a general principle of Co-existence, corresponding to Causation (the principle of Succession), we can only classify these uniformities as follows:

(*a*) The Geometrical; as that, in a four-sided figure, if the opposite angles are equal, the opposite sides are equal and parallel.—Countless similar uniformities of co-existence are disclosed by Geometry. The co-existent facts do not cause one another, nor are they jointly caused by something else; they are mutually involved: such is the nature of space.—

(*b*) Universal co-inherences among the properties of concrete things.—The chief example is the co-inherence of gravity with inertia in all material bodies. There is, I believe, no other entirely satisfactory case; but some good approximations to such uniformity are known to physical science.

(*c*) Co-existence due to Causation; such as the positions of objects in space at any time.—The houses of a town are where they are, because they were put there; and they remain in their place as long as no other causes arise strong enough to remove or destroy them. Similarly, the relative positions of rocks in geological strata, and of trees in a forest, are due to causes.

(*d*) The co-inherence of properties in Natural Kinds; which we call the constitution, defining characters, or specific nature of such things.—Oxygen, platinum, sulphur and the other elements; water, common salt, alcohol and other compounds; the various species of plants and animals: all these are known to us as different groups of co-inherent properties. It may be conjectured that these groupings of properties are also due to causation, and sometimes the causes can be traced: but very often the causes are still unknown; and, until resolved into their causes, they must be taken as necessary data in the investigation of nature. Laws of the co-inherence of the properties of

Kinds do not, like laws of causation, admit of methodical proof upon their own principles, but only by constancy in experience and statistical probability (c. xix, § 4).

(*e*) There are also a few cases in which properties co-exist in an unaccountable way, without being co-extensive with any one species, genus, or order: as most metals are whitish, and scarlet flowers are wanting in fragrance. (On this § 7, see Venn's *Empirical Logic*, c. 4.)

§ 8. Inasmuch as Axioms of Uniformity are ultimate truths, they cannot be deduced; and inasmuch as they are universal, no proof by experience can ever be adequate. The grounds of our belief in them seem to be these:

(1) Every inference takes for granted an order of Nature corresponding with it; and every attempt to explain the origin of anything assumes that it is the transformation of something else: so that uniformity of order and conservation of matter and energy are necessary presuppositions of reasoning.

(2) On the rise of philosophic reflection, these tacit presuppositions are first taken as dogmas, and later as postulates of scientific generalisation, and of the architectonic unification of science. Here they are indispensable.

(3) The presuppositions or postulates are, in some measure, verifiable in practical life and in scientific demonstration, and the better verifiable as our methods become more exact.

(4) There is a cause of this belief that cannot be said to contain any evidence for it, namely, the desire to find in Nature a foundation for confidence in our own power to foresee and to control events.

CHAPTER XIV

CAUSATION

§ 1. For the theory of Induction, the specially important aspect of the Uniformity of Nature is Causation.

For (1) the Principles of Contradiction and Excluded Middle are implied in all logical operations, and need no further explication.

(2) That one thing is a mark of another or constantly related to it, must be established by Induction; and the surest of all marks is a Cause. So that the application of the axiom of the Syllogism in particular cases requires, when most valid, a previous appeal to Causation.

(3) The uniformity of Space and of Time is involved in Causation, so far as we conceive Causation as essentially matter in motion—for motion is only known as a traversing of space in time; and so far as forces vary in any way according to the distance between bodies; so that if space and time were not uniform, causation would be irregular. Not that time and space are agents, but they are conditions of every agent's operation.

(4) The persistence of Matter and Energy, being nothing else than Causation in the general movement of the world, is applied under the name of that principle in explaining any particular limited phenomenon, such as a soap-bubble, or a thunderstorm, or the tide.

(5) As to co-existences, the Geometrical do not belong to Logic: those involved in the existence of plants, animals, and inorganic bodies, must, as far as possible, be traced to causes; and so, of course, must the relative positions of objects in space at any time: and what Co-existences remain do not admit of methodical inductive treatment; they will be briefly discussed in chap. xix.

Causation, then, is that mode or aspect of the Uniformity of Nature which especially concerns us as in Induction; and we must make it as definite as possible. It is nothing occult, but merely a convenient name for phenomena in a particular relation to other phenomena, called their effect. Similarly, if the word 'force' is sometimes used for convenience in analysing causation, it means nothing more than something in time and space, itself moving, or tending to move, or hindering or accelerating other things. If any one does not find these words convenient for the purpose, he can use others.

§ 2. A Cause, according to Mill, is "the invariable unconditional antecedent" of a given phenomenon. To enlarge upon this:

(1) A Cause is *relative to a given phenomenon*, called the Effect. Logic has no method for investigating the cause of the universe as a whole, but only of a part or epoch of it: we select from the infinite continuum of Nature any portion that is neither too large nor too small for a trained mind to comprehend. The magnitude of the phenomenon may be a matter of convenience. If the cause of disease in general be too wide a problem, can fevers be dealt with; or, if that be too much, is typhus within the reach of inquiry? In short, how much can we deal with accurately?

(2) The given phenomenon is always *an event*; that is to say, not a new thing (nothing is wholly new), but a change in something, or in the relative position of things. We may ask the cause of the phases of the moon, of the freezing of water, of the kindling of a match, of a deposit of chalk, of the differentiation of species. To inquire the cause of France being a republic, or Russia an autocracy, implies that these countries were once otherwise governed, or had no government: to inquire the cause of the earth being shaped like an orange, implies that the matter of the earth had once another shape.

(3) The Cause is *antecedent* to the Effect, which accordingly is often called its *consequent*. This is often misunderstood and sometimes disputed. It has been said that the meaning of 'cause' implies an 'effect,' so that until an effect occurs there can be no cause. But this is a blunder; for whilst the word 'cause' implies 'effect,' it also implies the relative futurity of the effect; and effect implies the relative priority of the cause. The connotation of the words, therefore, agrees well enough with Mill's doctrine. In fact, the danger is that any pair of contrasted words may suggest too strongly that the phenomena denoted are separate in Nature; whereas every natural process is continuous. If water, dripping from the roof wears away a stone, it fell on the roof as rain; the rain came from a condensing cloud; the cloud was driven by the wind from the sea, whence it exhaled; and so on. There is no known beginning to this, and no break in it. We may take any one of these changes, call it an effect, and ask for its cause; or call it a cause, and ask for its effect. There is not in Nature one set of things called causes and another called effects; but every change is both cause (or a condition) of the future and effect of the past; and whether we consider an event as the one or the other, depends upon the direction of our curiosity or interest.

Still, taking the event as effect, its cause is the antecedent process; or, taking it as a cause, its effect is the consequent process. This follows from the conception of causation as essentially motion; for that *motion takes time* is (from the way our perceptive powers grow) an ultimate intuition. But, for the same reason, there is no interval of time between cause and effect; since all the time is filled up with motion.

Nor must it be supposed that the whole cause is antecedent to the effect as a whole: for we often take the phenomenon on such a scale that minutes, days, years, ages, may elapse before we consider the cause as exhausted (*e.g.*, an earthquake, a battle, an expansion of credit, natural selection operating on a given variety); and all that time the effect has been accumulating. But we may further consider such a cause as made up of moments or minute factors, and the effect as made up of corresponding moments; and then the cause, taken in its moments, is antecedent throughout to the effect, taken in its corresponding moments.

(4) The Cause is the *invariable* antecedent of the effect; that is to say, whenever a given cause occurs it always has the same effect: in this, in fact, consists the Uniformity of Causation. Accordingly, not every antecedent of an event is its Cause: to assume that it is so, is the familiar fallacy of arguing '*post hoc ergo propter hoc*.' Every event has an infinite number of antecedents that have no ascertainable connection with it: if a picture falls from the wall in this room, there may have occurred, just previously, an earthquake in New Zealand, an explosion in a Japanese arsenal, a religious riot in India, a political assassination in Russia and a vote of censure in the House of Commons, besides millions of other less noticeable events, between none of which and the falling of the picture can any direct causation be detected; though, no doubt, they are all necessary occurrences in the general world-process, and remotely connected. The cause, however, was that a door slammed violently in the room above and shook the wall, and that the picture was heavy and the cord old and rotten. Even if two events invariably occur one after the other, as day follows night, or as the report follows the flash of a gun, they may not be cause and effect, though it is highly probable that they are closely connected by causation; and in each of these two examples the events are co-effects of a common cause, and may be regarded as elements of its total effect. Still, whilst it is not true that every antecedent, or that every invariable antecedent, of an event is its cause, the cause is conceived of as some change in certain conditions, or some state and process of things, such that should it exactly recur the same event would invariably follow. If we consider the antecedent state and process of things very widely or very minutely, it never does exactly recur; nor does the consequent. But the purpose of induction is to get as near the truth as possible within the limits set by our faculties of observation and calculation. Complex causal instances that are most unlikely to recur as a whole, may be analysed into the laws of their constituent conditions.

(5) The Cause is the Unconditional Antecedent. A cause is never simple, but may be analysed into several conditions; and 'Condition' means any necessary factor of a Cause: any thing or agent that exerts, absorbs, transforms, or deflects energy; or any relation of time or space in which agents stand to one another. A positive condition is one that cannot be omitted without frustrating the effect; a negative condition is one that cannot be introduced without frustrating the effect. In the falling of the picture, *e.g.*, the positive conditions were the picture (as being heavy), the slamming of the door, and the weakness of the cord: a negative condition was that the picture should have no support but the cord. When Mill, then, defines the Cause of any event as its "unconditional" antecedent, he means that it is that group of conditions (state and process of things) which, without any further condition, is followed by the event in question: it is the least antecedent that suffices, positive conditions being present and negative absent.

Whatever item of the antecedent can be left out, then, without affecting the event, is no part of the cause. Earthquakes have happened in New Zealand and votes of censure in the House of Commons without a picture's falling in this room: they were not unconditional antecedents; something else was needed to bring down a picture. Unconditionality also distinguishes a true cause from an invariable antecedent that is only a co-effect: for when day follows night something else happens; the Earth rotates upon her axis: a flash of gunpowder is not an unconditional antecedent of a report; the powder must be ignited in a closed chamber.

By common experience, and more precisely by experiment, it is found possible to select from among the antecedents of an event a certain number upon which, so far as can be perceived, it is dependent, and to neglect the rest: to purge the cause of all irrelevant antecedents is the great art of inductive method. Remote or minute conditions may indeed modify the event in ways so refined as to escape our notice. Subject to the limitations of our human faculties, however, we are able in many cases to secure an unconditional antecedent upon which a certain event invariably follows. Everybody takes this for granted: if the gas will not burn, or a gun will not go off, we wonder 'what can be wrong with it,' that is, what positive condition is wanting, or what negative one is present. No one now supposes that gunnery depends upon those "remotest of all causes," the stars, or upon the sun being in Sagittarius rather than in Aquarius, or that one shoots straightest with a silver bullet, or after saying the alphabet backwards.

(6) That the Cause of any event is an Immediate Antecedent follows from its being an unconditional one. For if there are three events, A B C, causally connected, it is plain that A is not the unconditional antecedent of C, but requires the further condition of first giving rise to B. But that is not all; for the B that gives rise to C is never merely the effect of A; it involves something further. Take such a simple case as the motion of the earth round the sun (neglecting all other conditions, the other planets, *etc.*); and let the earth's motion at three successive moments be A B C: A is not the whole cause of B in velocity and direction; we must add relation to the sun, say x. But then, again, the cause of C will not be merely Bx, for the relation to the sun will have altered; so that we must represent it as Bx'. The series, therefore, is Ax Bx' C. What is called a "remote cause" is, therefore, doubly conditional; first, because it supposes an intervening cause; and secondly, because it only in part determines the conditions that constitute this intervening cause.

The immediacy of a cause being implied in its unconditionalness, is an important clue to it; but as far as the detection of causes depends upon sense-perception, our powers (however aided by instruments) are unequal to the subtlety of Nature. Between the event and what seems to us the immediate antecedent many things (molecular or etherial changes) may happen in Chemistry or Physics. The progress of science would be impossible were not observation supplemented by hypothesis and calculation. And where phenomena are treated upon a large scale, as in the biological and social sciences, immediacy, as a mark of causation, must be liberally interpreted. So far, then, as to the qualitative character of Causation.

(7) But to complete our account of it, we must briefly consider its quantitative character. As to the Matter contained, and as to the Energy embodied, Cause and Effect are conceived to be *equal*. As to matter, indeed, they may be more properly called identical; since the effect is nothing but the cause redistributed. When oxygen combines with hydrogen to form water, or with mercury to form red precipitate, the weight of the compound is exactly equal to the weight of the elements combined in it; when a shell explodes and knocks down a wall, the materials of the shell and wall are scattered about. As to energy, we see that in the heavenly bodies, which meet with no sensible impediment, it remains the same from age to age: with things 'below the moon' we have to allow for the more or less rapid conversion of the visible motion of a mass into other forms of energy, such as sound and heat. But the right understanding of this point involves physical considerations of some difficulty, as to which the reader must refer to appropriate books, such as Balfour Stewart's on *The Conservation of Energy*.

The comprehension of the quantitative aspect of causation is greatly aided by Bain's analysis of any cause into a 'Moving or an Inciting Power' and a 'Collocation' of circumstances. When a demagogue by making a speech stirs up a mob to a riot, the speech is the moving or inciting power; the mob already in a state of smouldering passion, and a street convenient to be wrecked, are the collocation. When a small quantity of strychnine kills a man, the strychnine is the inciting power; the nature of his nervo-muscular system, apt to be thrown into spasms by that drug, and all the organs of his body dependent on that system, are the collocation. Now any one who thinks only of the speech, or the drug, in these cases, may express astonishment at the disproportion of cause and effect:

"What great events from trivial causes spring!"

But, remembering that the whole cause of the riot included the excited mob, every one sees that its muscular power is enough to wreck a street; and remembering that breathing depends upon the normal action of the intercostal muscles, it is plain that if this action is stopped by strychnine, a man must die. Again, a slight rise of temperature may be a sufficient inciting power to occasion extensive chemical changes in a collocation of elements otherwise stable; a spark is enough to explode a powder magazine. Hence, when sufficient energy to account for any effect cannot be found in the inciting power, or manifestly active condition, we must look for it in the collocation which is often supposed to be passive.

And that reminds us of another common misapprehension, namely, that in Nature some things are passive and others active: the distinction between 'agent' and 'patient.' This is a merely relative distinction: in Nature all things are active. To the eye some things seem at rest and others in motion; but we know that nothing is really at rest, that everything palpitates with molecular change, and whirls with the planet through space. Everything that is acted upon reacts according to its own nature: the quietest-looking object (say, a moss-covered stone), if we try to push or lift it, pushes or pulls us back, assuring us that 'action and reaction are equal and opposite.' 'Inertia' does not mean want of vigour, but may be metaphorically described as the inexpugnable resolve of everything to have its own way.

The equality of cause and effect defines and interprets the unconditionality of causation. The cause, we have seen, is that group of conditions which, without any further condition, is followed by a given event. But how is such a group to be conceived? Unquantified, it admits only of a general description: quantified, it must mean a group of conditions equal to the effect in mass and energy, the essence of the physical world. Apparently, a necessary conception of the human mind: for if a cause seem greater than its effect, we ask what has become of the surplus matter and energy; or if an effect seem greater than its cause, we ask whence the surplus matter and energy has arisen. So convinced of this truth is every experimenter, that if his results present any deviation from it, he always assumes that it is he who has made some mistake or oversight, never that there is indeterminism or discontinuity in Nature.

The transformation of matter and energy, then, is the essence of causation: because it is continuous, causation is immediate; and because in the same circumstances the transformation always follows the same course, a cause has invariably the same effect. If a fire be lit morning after morning in the same grate, with coal, wood, and paper of the same quality and similarly arranged, there will be each day the same flaming of paper, crackling of wood and glowing of coal, followed in about the same time by the same reduction of the whole mass partly to ashes and partly to gases and smoke that have gone up the chimney. The flaming, crackling and glowing are, physically, modes of energy; and the change of materials into gas and ashes is a chemical and physical redistribution: and, if some one be present, he will be aware of all this; and then, besides the physical changes, there will be sensations of light, sound and heat; and these again will be always the same in the same circumstances.

The Cause of any event, then, when exactly ascertainable, has five marks: it is (quantitatively) *equal* to the effect, and (qualitatively) *the immediate, unconditional, invariable antecedent of the effect.*

§ 3. This scientific conception of causation has been developed and rendered definite by the investigations of those physical sciences that can avail themselves of exact experiments and mathematical calculation; and it is there, in Chemistry and Physics, that it is most at home. The conception can indeed be carried into the Biological and Social Sciences, even in its quantitative form, by making the proper allowances. For the limbs of animals are levers, and act upon mechanical principles; and digestion and the aeration of the blood by breathing are partly chemical processes. There is a quantitative relation between the food a man eats and the amount of work he can do. The numbers of any species of plant or animal depend upon the food supply. The value of a country's imports is equal to the value of its exports and of the services it renders to foreigners. But, generally, the less experiment and exact calculation are practicable in any branch of inquiry, the less rigorously can the conception of causation be applied there, the more will its application depend upon the qualitative marks, and the more need there will be to use it judiciously. In every inquiry the greatest possible precision must be aimed at; but it is unreasonable to expect in any case more precise proof than the subject admits of in the existing state of culture.

Wherever mental action is involved, there is a special difficulty in applying the physical notion of causation. For if a Cause be conceived of as matter in motion, a thought, or feeling, or volition can be neither cause nor effect. And since mental action is involved in all social affairs, and in the life of all men and animals, it may seem impossible to interpret social or vital changes according to laws of causation. Still, animals and men are moving bodies; and it is recognised that their thoughts and feelings are so connected with their movements and with the movements of other things acting upon them, that we can judge of one case by another; although the connection is by no means well understood, and the best words (such as all can agree to use) have not yet been found to express even what we know about it. Hence, a regular connection being granted, I have not hesitated, to use biological and social events and the laws of them, to illustrate causation and induction; because, though less exact than chemical or mechanical examples, they are to most people more familiar and interesting.

In practical affairs, it is felt that everything depends upon causation; how to play the fiddle, or sail a yacht, or get one's living, or defeat the enemy. The price of pig-iron six months hence, the prospects of the harvest, the issue in a Coroner's Court, Home Rule and Socialism, are all questions of causation. But, in such cases, the conception of a cause is rarely applied in its full scientific acceptation, as the unconditional antecedent, or 'all the conditions' (neither more nor less) upon which the event depends. This is not because men of affairs are bad logicians, or incapable of scientific comprehension; for very often the reverse is conspicuously true; but because practical affairs call for promptitude and a decisive seizing upon what is predominantly important. How learn to play the fiddle? "Go to a good teacher." (Then, beginning young enough, with natural aptitude and great diligence, all may be well.) How defeat the enemy? "Be two to one at the critical juncture." (Then, if the men are brave, disciplined, well armed and well fed, there is a good chance of victory.) Will the price of iron improve? "Yes: for the market is oversold": (that is, many have sold iron who have none to deliver, and must at some time buy it back; and that will put up the price—if the stock is not too great, if the demand does not fall off, and if those who have bought what they cannot pay for are not in the meanwhile obliged to sell.) These prompt and decisive judgments (with the parenthetic considerations unexpressed) as to what is the Cause, or predominantly important condition, of any event, are not as good as a scientific estimate of all the conditions, when this can be obtained; but, when time is short, the insight of trained sagacity may be much better than an imperfect theoretical treatment of such problems.

§ 4. To regard the Effect of certain antecedents in a narrow selective way, is another common mistake. In the full scientific conception of an Effect it is the sum of the unconditional consequences of a given state and process of things: the consequences immediately flowing from that situation without further conditions. Always to take account of all the consequences of any cause would no doubt be impracticable; still the practical, as well as the scientific interest, often requires that we should enlarge our views of them; and there is no commoner error in private effort or in legislation than to aim at some obvious good, whilst overlooking other consequences of our action, the evil of which may far outweigh that good. An important consequence of eating is to satisfy hunger, and this is the ordinary motive to eat; but it is a poor account of the physiological consequences. An important consequence of firing a gun is the propulsion of the bullet or shell; but there are many other consequences in the whole effect, and one of them is the heating of the barrel, which, accumulating with rapid firing, may at last put the gun out of action. The tides have consequences to shipping and in the wear and tear of the coast that draw every one's attention; but we are told that they also retard the rotation of the earth, and at last may cause it to present always the same face to the sun, and, therefore, to be uninhabitable. Such concurrent consequences of any cause may be called its Co-effects: the Effect being the sum of them.

The neglect to take account of the whole effect (that is, of all the co-effects) in any case of causation is perhaps the reason why many philosophers have maintained the doctrine of a "Plurality of Causes": meaning not that more than one condition is operative in the antecedent of every event (which is true), but that the same event may be due at different times to different antecedents, that in fact there may be *vicarious* causes. If, however, we take any effect as a whole, this does not seem to be true. A fire may certainly be lit in many ways: with a match or a flint and steel, or by rubbing sticks together, or by a flash of lightning: have we not here a plurality of causes? Not if we take account of the whole effect; for then we shall find it modified in each case according to the difference of the cause. In one case there will be a burnt match, in another a warm flint, in the last a changed state of electrical tension. And similar differences are found in cases of death under different conditions, as stabbing, hanging, cholera; or of shipwreck from explosion, scuttling, tempest. Hence a Coroner's Court expects to find, by examining a corpse, the precise cause of death. In short, if we knew the facts minutely enough, it would be found that there is only one Cause (sum of conditions) for each Effect (sum of co-effects), and that the order of events is as uniform backwards as forwards.

Still, as we are far from knowing events minutely, it is necessary in practical affairs, and even in the more complex and unmanageable scientific investigations, especially those that deal with human life, to acknowledge a possible plurality of causes for any effect. Indeed, forgetfulness of this leads to many rash generalisations; as that 'revolutions always begin in hunger'; or that 'myths are a disease of language.' Then there is great waste of ingenuity in reconciling such propositions with the recalcitrant facts. A scientific method recognises that there may be other causes of effects thus vaguely conceived, and then proceeds to distinguish in each class of effects the peculiarities due to different causes.

§ 5. The understanding of the complex nature of Causes and Effects helps us to overcome some other difficulties that perplex the use of these words. We have seen that the true cause is an *immediate* antecedent; but if the cause is confounded with *one* of its constituent conditions, it may seem to have long preceded the event which is regarded as its effect. Thus, if one man's death is ascribed to another's desire of revenge, this desire may have been entertained for years before the assassination occurred: similarly, if a shipwreck is ascribed to a sunken reef, the rock was waiting for ages before the ship sailed that way. But, of course, neither the desire of revenge nor the sunken rock was 'the sum of the conditions' on which the one or the other event depended: as soon as this is complete the effect appears.

We have also seen the true effect of any state and process of things is the immediate consequence; but if the effect be confounded with *one* of its constituent factors, it may seem to long outlive the cessation of the cause. Thus, in nearly every process of human industry and art, one factor of the effect—a road, a house, a tool, a picture—may, and generally does, remain long after the work has ceased: but such a result is not the whole effect of the operations that produce it. The other factors may be, and some always are, evanescent. In most of such works some heat is produced by hammering or friction, and the labourers are fatigued; but these consequences soon pass off. Hence the effect as a whole only momentarily survives the cause. Consider a pendulum which, having been once set agoing, swings to and fro in an arc, under the joint control of the shaft, gravitation and its own inertia: at every moment its speed and direction change; and each change may be considered as an effect, of which the antecedent change was one condition. In such a case as this, which, though a very simple, is a perfectly fair example of all causation, the duration of either cause or effect is quite insensible: so that, as Dr. Venn says, an Effect, rigorously conceived, is only "the initial tendency" of its Cause.

§ 6. Mill contrasted two forms under which causation appears to us: that is to say, the conditions constituting a cause may be modified, or 'intermixed' in the effect, in two ways, which are typified respectively by Mechanical and Chemical action. In mechanical causation, which is found in Astronomy and all branches of Physics, the effects are all reducible to modes of energy, and are therefore commensurable with their causes. They are either directly commensurable, as in the cases treated in the consideration of the mechanical powers; or, if different forms of energy enter into cause and effect, such as mechanical energy, electrical energy, heat, these different forms are severally reducible to units, between which equivalents have been established. Hence Mill calls this the "homogeneous intermixture of effects," because the antecedents and consequents are fundamentally of the same kind.

In chemical causation, on the other hand, cause and effect (at least, as they present themselves to us) differ in almost every way: in the act of combination the properties of elements (except weight) disappear, and are superseded by others in the compound. If, for example, mercury (a heavy, silvery liquid) be heated in contact with oxygen (a colourless gas), oxide of mercury is formed (red precipitate, which is a powder). This compound presents very different phenomena from those of its elements; and hence Mill called this class of cases "the heteropathic intermixture of effects." Still, in chemical action, the effect is not (in Nature) heterogeneous with the cause: for the weight of a compound is equal to the sum of the weights of the elements that are merged in it; and an equivalence has been ascertained between the energy of chemical combination and the heat, light, *etc.,* produced in the act of combination.

The heteropathic intermixture of effects is also found in organic processes (which, indeed, are partly chemical): as when a man eats bread and milk, and by digestion and assimilation converts them into nerve, muscle and bone. Such phenomena may make us wonder that people should ever have believed that 'effects resemble their causes,' or that 'like produces like.' A dim recognition of the equivalence of cause and effect in respect of matter and motion may have aided the belief; and the resemblance of offspring to parents may have helped: but it is probably a residuum of magical rites; in which to whistle may be regarded as a means of raising the wind, because the wind whistles; and rain-wizards may make a victim shed tears that the clouds also may weep.

§ 7. Another consideration arises out of the complex character of causes and effects. When a cause consists of two or more conditions or forces, we may consider what effect any one of them would have if it operated alone, that is to say, its *Tendency*. This is best illustrated by the Parallelogram of Forces: if two forces acting upon a point, but not in the same direction, be represented by straight lines drawn in the direction of the forces, and in length proportional to their magnitudes, these lines, meeting in an angle, represent severally the tendencies of the forces; whilst if the parallelogram be completed on these lines, the diagonal drawn from the point in which they meet represents their *Resultant* or effect.

Again, considering the tendency of any force if it operated alone, we may say that, when combined with another force (not in the same direction) in any resultant, its tendency is *counteracted*: either partially, when the direction of the resultant is different; or wholly when, the other force being equal and opposite, the resultant is equilibrium. If the two forces be in the same direction, they are merely added together. Counteraction is only one mode of combination; in no case is any force destroyed.

Sometimes the separate tendencies of combined forces can only be theoretically distinguished: as when the motion of a projectile is analysed into a tendency to travel in the straight line of its discharge, and a tendency to fall straight to the ground. But sometimes a tendency can be isolated: as when,—after dropping a feather in some place sheltered from the wind, and watching it drift to and fro, as the air, offering unequal resistances to its uneven surface, counteracts its weight with varying success, until it slowly settles upon the ground,—we take it up and drop it again in a vacuum, when it falls like lead. Here we have the tendency of a certain cause (namely, the relation between the feather and the earth) free from counteraction: and this is called the *Elimination* of the counteracting circumstances. In this case indeed there is physical elimination; whereas, in the case of a projectile, when we say that its actual motion is resolvable (neglecting the resistance of the air) into two tendencies, one in the line of discharge, the other earthwards, there is only theoretical elimination of either tendency, considered as counteracting the other; and this is more specifically called the *Resolution* or Analysis of the total effect into its component conditions. Now, Elimination and Resolution may be said to be the essential process of Induction in the widest sense of the term, as including the combination of Induction with Deduction.

The several conditions constituting any cause, then, by aiding or counteracting one another's tendencies, jointly determine the total effect. Hence, viewed in relation one to another, they may be said to stand in *Reciprocity* or mutual influence. This relation at any moment is itself one of co-existence, though it is conceived with reference to a possible effect. As Kant says, all substances, as perceived in space at the same time, are in reciprocal activity. And what is true of the world of things at any moment (as connected, say, by gravity), is true of any selected group of circumstances which we regard as the particular cause of any event to come. The use of the concept of reciprocity, then, lies in the analysis of a cause: we must not think of reciprocity as obtaining in the succession of cause and effect, as if the effect could turn back upon its cause; for as the effect arises its cause disappears, and is irrecoverable by Nature or Magic. There are many cases of rhythmic change and of moving equilibria, in which one movement or process produces another, and this produces something closely resembling the former, and so on in long series; as with the swing of a pendulum or the orbit of a planet: but these are series of cause and effect, not of reciprocity.

CHAPTER XV
INDUCTIVE METHOD

§ 1. It is necessary to describe briefly the process of investigating laws of causation, not with the notion of teaching any one the Art of Discovery, which each man pursues for himself according to his natural gifts and his experience in the methods of his own science, but merely to cast some light upon the contents of the next few chapters. Logic is here treated as a process of proof; proof supposes that some general proposition or hypothesis has been suggested as requiring proof; and the search for such propositions may spring from scientific curiosity or from practical interests.

We may, as Bain observes (*Logic*: B. iii. ch. 5), desire to detect a process of causation either (1) amidst circumstances that have no influence upon the process but only obscure it; as when, being pleased with a certain scent in a garden, we wish to know from what flower it rises; or, being attracted by the sound of some instrument in an orchestra, we desire to know which it is: or (2) amidst circumstances that alter the effect from what it would have been by the sole operation of some cause; as when the air deflects a falling feather; or in some more complex case, such as a rise or fall of prices that may extend over many years.

To begin with, we must form definite ideas as to what the phenomenon is that we are about to investigate; and in a case of any complexity this is best done by writing a detailed description of it: *e.g.*, to investigate the cause of a recent fall of prices, we must describe exactly the course of the phenomenon, dating the period over which it extends, recording the successive fluctuations of prices, with their maxima and minima, and noting the classes of goods or securities that were more or less affected, *etc.*

Then the first step of elimination (as Bain further observes) is "to analyse the situation mentally," in the light of analogies suggested by our experience or previous knowledge. Dew, for example, is moisture formed upon the surface of bodies from no apparent source. But two possible sources are easily suggested by common experience: is it deposited from the air, like the moisture upon a mirror when we breathe upon it; or does it exude from the bodies themselves, like gum or turpentine? Or, again, as to a fall of prices, a little experience in business, or knowledge of Economics, readily suggests two possible explanations: either cheaper production in making goods or carrying them; or a scarcity of that in which the purchasing power of the chief commercial nations is directly expressed, namely, gold.

Having thus analysed the situation and considered the possibility of one, two, three, or more possible causes, we fix upon one of them for further investigation; that is to say, we frame an hypothesis that this is the cause. When an effect is given to find its cause, an inquirer nearly always begins his investigations by thus framing an hypothesis as to the cause.

The next step is to try to *verify* this Hypothesis. This we may sometimes do by *varying the circumstances* of the phenomenon, according to the Canons of direct Inductive Proof to be discussed in the next chapter; that is to say, by *observing* or *experimenting* in such a way as to get rid of or eliminate the obscuring or disturbing conditions. Thus, to find out which flower in a garden gives a certain scent, it is usually enough to rely on observation, going up to the likely flowers one after the other and smelling them: at close quarters, the greater relative intensity of the scent is sufficiently decisive. Or we may resort to a sort of experiment, plucking a likely flower, as to which we frame the hypothesis (this is the cause), and carrying it to some place where the air is free from conflicting odours. Should observation or experiment disprove our first hypothesis we try a second; and so on until we succeed, or exhaust the known possibilities.

But if the phenomenon is so complex and extensive as a continuous fall of prices, direct observation or experiment is a useless or impossible method; and we must then resort to Deduction; that is, to indirect Induction. If, for example, we take the hypothesis that the fall is due to a scarcity of gold, we must show that there is a scarcity; what effect such a scarcity may be expected to have upon prices from the acknowledged laws of prices, and from the analogy of other cases of an expanded or restricted currency; that this expectation agrees with the statistics of recent commerce: and finally, that the alternative hypothesis that the fall is due to cheaper production is not true; either because there has not been a sufficient cheapening of general production; or because, if there has been, the results to be rationally expected from it are not such as to agree with the statistics of recent commerce. (Ch. xviii.)

But now suppose that, a phenomenon having been suggested for explanation, we are unable at the time to think of any cause—to frame any hypothesis about it; we must then wait for the phenomenon to occur again, and, once more observing its course and accompaniments and trying to recall its antecedents, do our best to conceive an hypothesis, and proceed as before. Thus, in the first great epidemic of influenza, some doctors traced it to a deluge in China, others to a volcanic eruption near Java; some thought it a mild form of Asiatic plague, and others caught a specific microbe. As the disease often recurred, there were fresh opportunities of framing hypotheses; and the microbe was identified.

Again, the investigation may take a different form: given a supposed Cause to find its Effect; *e.g.*, a new chemical element, to find what compounds it forms with other elements; or, the spots on the sun—have they any influence upon our weather?

Here, if the given cause be under control, as a new element may be, it is possible to try experiments with it according to the Canons of Inductive Proof. The inquirer may form some hypothesis or expectation as to the effects, to guide his observation of them, but will be careful not to hold his expectation so confidently as to falsify his observation of what actually happens.

But if the cause be, like the sun-spots, not under control, the inquirer will watch on all sides what events follow their appearance and development; he must watch for consequences of the new cause he is studying in many different circumstances, that his observations may satisfy the canons of proof. But he will also resort for guidance to deduction; arguing from the nature of the cause, if anything is known of its nature, what consequences may be expected, and comparing the results of this deduction with any consequent which he suspects to be connected with the cause. And if the results of deduction and observation agree, he will still consider whether the facts observed may not be due to some other cause.

A cause, however, may be under control and yet be too dangerous to experiment with; such as the effects of a poison—though, if too dangerous to experiment with upon man, it may be tried upon animals; or such as a proposed change of the constitution by legislation; or even some minor Act of Parliament, for altering the Poor Law, or regulating the hours of labour. Here the first step must be deductive. We must ask what consequences are to be expected from the nature of the change (comparing it with similar changes), and from the laws of the special circumstances in which it is to operate? And sometimes we may partially verify our deduction by trying experiments upon a small scale or in a mild form. There are conflicting deductions as to the probable effect of giving Home Rule to Ireland; and experiments have been made in more or less similar cases, as in the Colonies and in some foreign countries. As to the proposal to make eight hours the legal limit of a day's labour in all trades, we have all tried to forecast the consequences of this; and by way of verification we might begin with nine hours; or we might induce some other country to try the experiment first. Still, no verification by experiments on a small scale, or in a mild form, or in somewhat similar yet different circumstances, can be considered logically conclusive. What proofs are conclusive we shall see in the following chapters.

§ 2. To begin with the conditions of direct Induction.—An Induction is an universal real proposition, based on observation, in reliance on the uniformity of Nature: when well ascertained, it is called a Law. Thus, that all life depends on the presence of oxygen is (1) an universal proposition; (2) a real one, since the 'presence of oxygen' is not connoted by 'life'; (3) it is based on observation; (4) it relies on the uniformity of Nature, since all cases of life have not been examined.

Such a proposition is here called 'an induction,' when it is inductively proved; that is, proved by facts, not merely deduced from more general premises (except the premise of Nature's uniformity): and by the 'process of induction' is meant the method of inductive proof. The phrase 'process of induction' is often used in another sense, namely for the inference or judgment by which such propositions are arrived at. But it is better to call this 'the process of hypothesis,' and to regard it as a preliminary to the process of induction (that is, proof), as furnishing the hypothesis which, if it can stand the proper tests, becomes an induction or law.

§ 3. Inductive proofs are usually classed as Perfect and Imperfect. They are said to be perfect when all the instances within the scope of the given proposition have been severally examined, and the proposition has been found true in each case. But we have seen (chap. xiii. § 2) that the instances included in universal propositions concerning Causes and Kinds cannot be exhaustively examined: we do not know all planets, all heat, all liquids, all life, *etc.*; and we never can, since a man's life is never long enough. It is only where the conditions of time, place, etc., are arbitrarily limited that examination can be exhaustive. Perfect induction might show (say) that every member of the present House of Commons has two Christian names. Such an argument is sometimes exhibited as a Syllogism in Darapti with a Minor premise in U., which legitimates a Conclusion in A., thus:

A.B. to Z have two Christian names; A.B. to Z are all the present M.P.'s: ∴ All the present M.P.'s have two Christian names.

But in such an investigation there is no need of logical method to find the major premise; it is mere counting: and to carry out the syllogism is a hollow formality. Accordingly, our definition of Induction excludes the kind unfortunately called Perfect, by including in the notion of Induction a reliance on the uniformity of Nature; for this would be superfluous if every instance in question had been severally examined. Imperfect Induction, then, is what we have to deal with: the method of showing the credibility of an universal real proposition by an examination of *some* of the instances it includes, generally a small fraction of them.

§ 4. Imperfect Induction is either Methodical or Immethodical. Now, Method is procedure upon a principle; and if the method is to be precise and conclusive, the principle must be clear and definite.

There is a Geometrical Method, because the axioms of Geometry are clear and definite, and by their means, with the aid of definitions, laws are deduced of the equality of lines and angles and other relations of position and magnitude in space. The process of proof is purely Deductive (the axioms and definitions being granted). Diagrams are used not as facts for observation, but merely to fix our attention in following the general argument; so that it matters little how badly they are drawn, as long as their divergence from the conditions of the proposition to be proved is not distracting. Even the appeal to "superposition" to prove the equality of magnitudes (as in Euclid I. 4), is not an appeal to observation, but to our judgment of what is implied in the foregoing conditions. Hence no inference is required from the special case to all similar ones; for they are all proved at once.

There is also, as we have seen, a method of Deductive Logic resting on the Principles of Consistency and the *Dictum de omni et nullo*. And we shall find that there is a method of Inductive Logic, resting on the principle of Causation.

But there are a good many general propositions, more or less trustworthy within a certain range of conditions, which cannot be methodically proved for want of a precise principle by which they may be tested; and they, therefore, depend upon Immethodical Induction, that is, upon the examination of as many instances as can be found, relying for the rest upon the undefinable principle of the Uniformity of Nature, since we are not able to connect them with any of its definite modes enumerated in chap. xiii. § 7. To this subject we shall return in chap. xix., after treating of Methodical Induction, or the means of determining that a relation of events is of the nature of cause and effect, because the relation can be shown to have the marks of causation, or some of them.

§ 5. Observations and Experiments are the *material* grounds of Induction. An experiment is an observation made under prepared, and therefore known, conditions; and, when obtainable, it is much to be preferred. Simple observation shows that the burning of the fire depends, for one thing, on the supply of air; but it cannot show us that it depends on oxygen. To prove this we must make experiments as by obtaining pure oxygen and pure nitrogen (which, mixed in the proportion of one to four, form the air) in separate vessels, and then plunging a burning taper into the oxygen—when it will blaze fiercely; and again plunging it into the nitrogen—when it will be extinguished. This shows that the greater part of the air does nothing to keep the fire alight, except by diminishing its intensity and so making it last longer. Experiments are more perfect the more carefully they are prepared, and the more completely the conditions are known under which the given phenomenon is to be observed. Therefore, they become possible only when some knowledge has already been gained by observation; for else the preparation which they require could not be made.

Observation, then, was the first material ground of Induction, and in some sciences it remains the chief ground. The heavenly bodies, the winds and tides, the strata of the earth, and the movements of history, are beyond our power to experiment with. Experiments upon the living body or mind are indeed resorted to when practicable, even in the case of man, as now in all departments of Psychology; but, if of a grave nature, they are usually thought unjustifiable. And in political affairs experiments are hindered by the reflection, that those whose interests are affected must bear the consequences and may resent them. Hence, it is in physical and chemical inquiries and in the physiology of plants and animals (under certain conditions) that direct experiment is most constantly practised.

Where direct experiment is possible, however, it has many advantages over unaided observation. If one experiment does not enable us to observe the phenomenon satisfactorily, we may try again and again; whereas the mere observer, who wishes to study the bright spots on Mars, or a commercial crisis, must wait for a favourable opportunity. Again, in making experiments we can vary the conditions of the phenomenon, so as to observe its different behaviour in each case; whereas he who depends solely on observation must trust the bounty of nature to supply him with a suitable diversity of instances. It is a particular advantage of experiment that a phenomenon may sometimes be 'isolated,' that is, removed from the influence of all agents except that whose operation we desire to observe, or except those whose operation is already known: whereas a simple observer, who has no control over the conditions of the subject he studies, can never be quite sure that its movements or changes are not due to causes that have never been conspicuous enough to draw his attention. Finally, experiment enables us to observe coolly and circumspectly and to be precise as to what happens, the time of its occurrence, the order of successive events, their duration, intensity and extent.

But whether we proceed by observation or experiment, the utmost attainable exactness of measurements and calculation is requisite; and these presuppose some Unit, in multiples or divisions of which the result may be expressed. This unit cannot be an abstract number as in Arithmetic, but must be one something—an hour, or a yard, or a pound—according to the nature of the phenomenon to be measured. But what is an hour, or a yard or a pound? There must in each case be some constant Standard of reference to give assurance that the unit may always have the same value. "The English pound is defined by a certain lump of platinum preserved at Westminster." The unit may be identical with the standard or some division or multiple of it; and, in measuring the same kind of phenomena, different units may be used for different purposes as long as each bears a constant relation to the standard. Thus, taking the rotation of the earth as the standard of Time, the convenient unit for long periods is a year (which is a multiple); for shorter periods, a day (which is identical); for shorter still, an hour (which is a division), or a second, or a thousandth of a second. (See Jevons' *Principles of Science*, ch. 14.)

§ 6. The principle of Causation is the *formal* ground of Induction; and the Inductive Canons derived from it are means of testing the formal sufficiency of observations to justify the statement of a Law. If we can observe the process of cause and effect in nature we may generalise our observation into a law, because that process is invariable. First, then, can we observe the course of cause and effect? Our power to do so is limited by the refinement of our senses aided by instruments, such as lenses, thermometers, balances, *etc*. If the causal process is essentially molecular change, as in the maintenance of combustion by oxygen, we cannot directly observe it; if the process is partly cerebral or mental, as in social movements which depend on feeling and opinion, it can but remotely be inferred; even if the process is a collision of moving masses (billiard-balls), we cannot really observe what happens, the elastic yielding, and recoil and the internal changes that result; though no doubt photography will throw some light upon this, as it has done upon the galloping of horses and the impact of projectiles. Direct observation is limited to the effect which any change in a phenomenon (or its index) produces upon our senses; and what we believe to be the causal process is a matter of inference and calculation. The meagre and abstract outlines of Inductive Logic are apt to foster the notion, that the evidence on which Science rests is simple; but it is amazingly intricate and cumulative.

Secondly, so far as we can observe the process of nature, how shall we judge whether a true causal instance, a relation of cause and effect, is before us? By looking for the five marks of Causation. Thus, in the experiment above described, showing that oxygen supports combustion, we find—(1) that the taper which only glowed before being plunged into the oxygen, bursts into flame when there—Sequence; (2) that this begins to happen at once without perceptible interval—Immediacy; (3) that no other agent or disturbing circumstance was present (the preparation of the experiment having excluded any such thing)—Unconditionalness; (4) the experiment may be repeated as often as we like with the same result—Invariableness. Invariableness, indeed, I do not regard as formally necessary to be shown, supposing the other marks to be clear; for it can only be proved within our experience; and the very object of Induction is to find grounds of belief beyond actual experience. However, for material assurance, to guard against his own liability to error, the inquirer will of course repeat his experiments.

The above four are the qualitative marks of Causation: the fifth and quantitative mark is the Equality of Cause and Effect; and this, in the above example, the Chemist determines by showing that, instead of the oxygen and wax that have disappeared during combustion, an equivalent weight of carbon dioxide, water, *etc.*, has been formed.

Here, then, we have all the marks of causation; but in the ordinary judgments of life, in history, politics, criticism, business, we must not expect such clear and direct proofs; in subsequent chapters it will appear how different kinds of evidence are combined in different departments of investigation.

§ 7. The Inductive Canons, to be explained in the next chapter, describe the character of observations and experiments that justify us in drawing conclusions about causation; and, as we have mentioned, they are derived from the principle of Causation itself. According to that principle, cause and effect are invariably, immediately and unconditionally antecedent and consequent, and are equal as to the matter and energy embodied.

Invariability can only be observed, in any of the methods of induction, by collecting more and more instances, or repeating experiments. Of course it can never be exhaustively observed.

Immediacy, too, in direct Induction, is a matter for observation the most exact that is possible.

Succession, or the relation itself of antecedent and consequent, must either be directly observed (or some index of it); or else ascertained by showing that energy gained by one phenomenon has been lost by another, for this implies succession.

But to determine the unconditionality of causation, or the indispensability of some condition, is the great object of the methods, and for that purpose the meaning of unconditionality may be further explicated by the following rules for the determination of a Cause.

A. Qualitative Determination

I.—For Positive Instances.

To prove a supposed Cause: (*a*) Any agent whose introduction among certain conditions (without further change) is followed by a given phenomenon; or, (*b*) whose removal is followed by the cessation (or modification) of that phenomenon, is (so far) the cause or an indispensable condition of it.

To find the Effect: (*c*) Any event that follows a given phenomenon, when there is no further change; or, (*d*) that does not occur when the conditions of a former occurrence are exactly the same, except for the absence of that phenomenon, is the effect of it (or is dependent on it).

II.—For Negative Instances.

To exclude a supposed Cause: (*a*) Any agent that can be introduced among certain conditions without being followed by a given phenomenon (or that is found without that phenomenon); or (*b*) that can be removed when that phenomenon is present without impairing it (or that is absent when that phenomenon is present), is not the cause, or does not complete the cause, of that phenomenon in those circumstances.

To exclude a supposed Effect: (*c*) Any event that occurs without the introduction (or presence) of a given phenomenon; or (*d*) that does not occur when that phenomenon is introduced (or is present), is not the effect of that phenomenon.

Subject to the conditions thus stated, the rules may be briefly put as follows:
I. (*a*) That which (without further change) is followed by a given event is its cause.
II. (*a*) That which is not so followed is not the cause.
I. (*b*) That which cannot be left out without impairing a phenomenon is a condition of it.
II. (*b*) That which can be left out is not a condition of it.

B. Quantitative Determination

The Equality of Cause and Effect may be further explained by these rules:
III. (*a*) When a cause (or effect) increases or decreases, so does its effect (or cause).
III. (*b*) If two phenomena, having the other marks of cause and effect, seem unequal, the less contains an unexplored factor.
III. (*c*) If an antecedent and consequent do not increase or decrease correspondingly, they are not cause and effect, so far as they vary.

It will next be shown that these propositions are variously combined in Mill's five Canons of Induction: Agreement, the Joint Method, Difference, Variations, Residues. The first three are sometimes called Qualitative Methods, and the two last Quantitative; and although this grouping is not quite accurate, seeing that Difference is often used quantitatively, yet it draws attention to an important distinction between a mere description of conditions and determination by exact measurement.

To avoid certain misunderstandings, some slight alterations have been made in the wording of the Canons. It may seem questionable whether the Canons add anything to the above propositions: I think they do. They are not discussed in the ensuing chapter merely out of reverence for Mill, or regard for a nascent tradition; but because, as describing the character of observations and experiments that justify us in drawing conclusions about causation, they are guides to the analysis of observations and to the preparation of experiments. To many eminent investigators the Canons (as such) have been unknown; but they prepared their work effectively so far only as they had definite ideas to the same purport. A definite conception of the conditions of proof is the necessary antecedent of whatever preparations may be made for proving anything.

CHAPTER XVI
THE CANONS OF DIRECT INDUCTION

§ 1. Let me begin by borrowing an example from Bain (*Logic*: B. III. c. 6). The North-East wind is generally detested in this country: as long as it blows few people feel at their best. Occasional well-known causes of a wind being injurious are violence, excessive heat or cold, excessive dryness or moisture, electrical condition, the being laden with dust or exhalations. Let the hypothesis be that the last is the cause of the North-East wind's unwholesome quality; since we know it is a ground current setting from the pole toward the equator and bent westward by the rotation of the earth; so that, reaching us over thousands of miles of land, it may well be fraught with dust, effluvia, and microbes. Now, examining many cases of North-East wind, we find that this is the only circumstance in which all the instances agree: for it is sometimes cold, sometimes hot; generally dry, but sometimes wet; sometimes light, sometimes violent, and of all electrical conditions. Each of the other circumstances, then, can be omitted without the N.E. wind ceasing to be noxious; but one circumstance is never absent, namely, that it is a ground current. That circumstance, therefore, is probably the cause of its injuriousness. This case illustrates:—

(I) The Canon of Agreement.

If two or more instances of a phenomenon under investigation have only one other circumstance (antecedent or consequent) in common, that circumstance is probably the cause (or an indispensable condition) or the effect of the phenomenon, or is connected with it by causation.

This rule of proof (so far as it is used to establish direct causation) depends, first, upon observation of an invariable connection between the given phenomenon and one other circumstance; and, secondly, upon I. (*a*) and II. (*b*) among the propositions obtained from the unconditionality of causation at the close of the last chapter.

To prove that A is causally related to *p*, suppose two instances of the occurrence of A, an antecedent, and *p*, a consequent, with concomitant facts or events—and let us represent them thus:

Antecedents: A B C   A D E

Consequents: p q r   p s t;

and suppose further that, in this case, the immediate succession of events can be observed. Then A is probably the cause, or an indispensable condition, of *p*. For, as far as our instances go, A is the invariable antecedent of *p*; and *p* is the invariable consequent of A. But the two instances of A or *p* agree in no other circumstance. Therefore A is (or completes) the unconditional antecedent of *p*. For B and C are not indispensable conditions of *p*, being absent in the second instance (Rule II. (*b*)); nor are D and E, being absent in the first instance. Moreover, *q* and *r* are not effects of A, being absent in the second instance (Rule II. (*d*)); nor are *s* and *t*, being absent in the first instance.

It should be observed that the cogency of the proof depends entirely upon its tending to show the unconditionality of the sequence A-*p*, or the indispensability of A as a condition of *p*. That *p* follows A, even immediately, is nothing by itself: if a man sits down to study and, on the instant, a hand-organ begins under his window, he must not infer malice in the musician: thousands of things follow one another every moment without traceable connection; and this we call 'accidental.' Even invariable sequence is not enough to prove direct causation; for, in our experience does not night invariable follow day? The proof requires that the instances be such as to show not merely what events *are* in invariable sequence, but also what *are not*. From among the occasional antecedents of *p* (or consequents of A) we have to eliminate the accidental ones. And this is done by finding or making 'negative instances' in respect of each of them. Thus the instance

A B C

p s t

is a negative instance of B and C considered as supposable causes of *p* (and of *q* and *r* as supposable effects of A); for it shows that they are absent when *p* (or A) is present.

To insist upon the cogency of 'negative instances' was Bacon's great contribution to Inductive Logic. If we neglect them, and merely collect examples of the sequence A-*p*, this is 'simple enumeration'; and although simple enumeration, when the instances of agreement are numerous enough, may give rise to a strong belief in the connection of phenomena, yet it can never be a methodical or logical proof of causation, since it does not indicate the unconditionalness of the sequence. For simple enumeration of the sequence A-*p* leaves open the possibility that, besides A, there is always some other antecedent of *p*, say X; and then X may be the cause of *p*. To disprove it, we must find, or make, a negative instance of X—where *p* occurs, but X is absent.

So far as we recognise the possibility of a plurality of causes, this method of Agreement cannot be quite satisfactory. For then, in such instances as the above, although D is absent in the first, and B in the second, it does not follow that they are not the causes of *p*; for they may be alternative causes: B may have produced *p* in the first instance, and D in the second; A being in both cases an accidental circumstance in relation to *p*. To remedy this shortcoming by the method of Agreement itself, the only course is to find more instances of *p*. We may never find a negative instance of A; and, if not, the probability that A is the cause of *p* increases with the number of instances. But if there be no antecedent that we cannot sometimes exclude, yet the collection of instances will probably give at last all the causes of *p*; and by finding the proportion of instances in which A, B, or X precedes *p*, we may estimate the probability of any one of them being the cause of *p* in any given case of its occurrence.

But this is not enough. Since there cannot really be vicarious causes, we must define the effect (*p*) more strictly, and examine the cases to find whether there may not be varieties of *p*, with each of which one of the apparent causes is correlated: A with p^1, B with p^{11}, X with p^{111}.

Or, again, it may be that none of the recognised antecedents is effective: as we here depend solely on observation, the true conditions may be so recondite and disguised by other phenomena as to have escaped our scrutiny. This may happen even when we suppose that the chief condition has been isolated: the drinking of foul water was long believed to cause dysentery, because it was a frequent antecedent; whilst observation had overlooked the bacillus, which was the indispensable condition.

Again, though we have assumed that, in the instances supposed above, immediate sequence is observable, yet in many cases it may not be so, if we rely only on the canon of Agreement; if instances cannot be obtained by experiment, and we have to depend on observation. The phenomena may then be so mixed together that A and p seem to be merely concomitant; so that, though connection of some sort may be rendered highly probable, we may not be able to say which is cause and which is effect. We must then try (as Bain says) to trace the expenditure of energy: if p gains when A loses, the course of events if from A to p.

Moreover, where succession cannot be traced, the method of Agreement may point to a connection between two or more facts (perhaps as co-effects of a remote cause) where direct causation seems to be out of the question: *e.g.*, that Negroes, though of different tribes, different localities, customs, *etc.*, are prognathous, woolly-haired and dolichocephalic.

The Method of Agreement, then, cannot by itself prove causation. Its chief use (as Mill says) is to suggest hypotheses as to the cause; which must then be used (if possible) experimentally to try if it produces the given effect. A bacillus, for example, being always found with a certain disease, is probably the chief condition of it: give it to a guinea-pig, and observe whether the disease appears in that animal.

Men often use arguments which, if they knew it, might be shown to conform more or less to this canon; for they collect many instances to show that two events are connected; but usually neglect to bring out the negative side of the proof; so that their arguments only amount to simple enumeration. Thus Ascham in his *Toxophilus*, insisting on the national importance of archery, argues that victory has always depended on superiority in shooting; and, to prove it, he shows how the Parthians checked the Romans, Sesostris conquered a great part of the known world, Tiberius overcame Arminius, the Turks established their empire, and the English defeated the French (with many like examples)—all by superior archery. But having cited these cases to his purpose, he is content; whereas he might have greatly strengthened his proof by showing how one or the other instance excludes other possible causes of success. Thus: the cause was not discipline, for the Romans were better disciplined than the Parthians; nor yet the boasted superiority of a northern habitat, for Sesostris issued from the south; nor better manhood, for here the Germans probably had the advantage of the Romans; nor superior civilisation, for the Turks were less civilised than most of those they conquered; nor numbers, nor even a good cause, for the French were more numerous than the English, and were shamefully attacked by Henry V. on their own soil. Many an argument from simple enumeration may thus be turned into an induction of greater plausibility according to the Canon of Agreement.

Still, in the above case, the effect (victory) is so vaguely conceived, that a plurality of causes must be allowed for: although, *e.g.*, discipline did not enable the Romans to conquer the Parthians, it may have been their chief advantage over the Germans; and it was certainly important to the English under Henry V. in their war with the French.

Here is another argument, somewhat similar to the above, put forward by H. Spencer with a full consciousness of its logical character. States that make war their chief object, he says, assume a certain type of organisation, involving the growth of the warrior class and the treatment of labourers as existing solely to sustain the warriors; the complete subordination of individuals to the will of the despotic soldier-king, their property, liberty and life being at the service of the State; the regimentation of society not only for military but also for civil purposes; the suppression of all private associations, *etc.* This is the case in Dahomey and in Russia, and it was so at Sparta, in Egypt, and in the empire of the Yncas. But the similarity of organisation in these States cannot have been due to race, for they are all of different races; nor to size, for some are small, some large; nor to climate or other circumstances of habitat, for here again they differ widely: the one thing they have in common is the military purpose; and this, therefore, must be the cause of their similar organisation. (*Political Institutions.*)

By this method, then, to prove that one thing is causally connected with another, say A with p, we show, first, that in all instances of p, A is present; and, secondly, that any other supposable cause of p may be absent without disturbing p. We next come to a method the use of which greatly strengthens the foregoing, by showing that where p is absent A is also absent, and (if possible) that A is the only supposable cause that is always absent along with p.

§ 2. The Canon of the Joint Method of Agreement in Presence and in Absence.

If (1) two or more instances in which a phenomenon occurs have only one other circumstance (antecedent or consequent) in common, while (2) two or more instances in which it does not occur (though in important points they resemble the former set of instances) have nothing else in common save the absence of that circumstance—the circumstance in which alone the two sets of instances differ throughout (being present in the first set and absent in the second) is probably the effect, or the cause, or an indispensable condition of the phenomenon.

The first clause of this Canon is the same as that of the method of Agreement, and its significance depends upon the same propositions concerning causation. The second clause, relating to instances in which the phenomenon is absent, depends for its probative force upon Prop. II. (*a*), and I. (*b*): its function is to exclude certain circumstances (whose nature or manner of occurrence gives them some claim to consideration) from the list of possible causes (or effects) of the phenomenon investigated. It might have been better to state this second clause separately as the Canon of the Method of Exclusions.

To prove that A is causally related to p, let the two sets of instances be represented as follows:

Instances of Presence.			Instances of Absence.		
A	B	C	C	H	F
p	q	r	r	x	v
A	D	E	B	D	K
p	s	t	q	y	s

A	F	G		E	G	M
p	*u*	*v*		*t*	*f*	*u*

Then A is probably the cause or a condition of *p*, or *p* is dependent upon A: first, by the Canon of Agreement in Presence, as represented by the first set of instances; and, secondly, by Agreement in Absence in the second set of instances. For there we see that C, H, F, B, D, K, E, G, M occur without the phenomenon *p*, and therefore (by Prop. II. (*a*)) are not its cause, or not the whole cause, unless they have been counteracted (which is a point for further investigation). We also see that *r, v, q, s, t, u* occur without A, and therefore are not the effects of A. And, further, if the negative instances represent all possible cases, we see that (according to Prop. I. (*b*)) A is the cause of *p*, because it cannot be omitted without the cessation of *p*. The inference that A and *p* are cause and effect, suggested by their being present throughout the first set of instances, is therefore strengthened by their being both absent throughout the second set.

So far as this Double Method, like the Single Method of Agreement, relies on observation, sequence may not be perceptible in the instances observed, and then, direct causation cannot be proved by it, but only the probability of causal connection; and, again, the real cause, though present, may be so obscure as to evade observation. It has, however, one peculiar advantage, namely, that if the second list of instances (in which the phenomenon and its supposed antecedent are both absent) can be made exhaustive, it precludes any hypothesis of a plurality of causes; since all possible antecedents will have been included in this list without producing the phenomenon. Thus, in the above symbolic example, taking the first set of instances, the supposition is left open that B, C, D, E, F, G may, at one time or another, have been a condition of *p*; but, in the second list, these antecedents all occur, here or there, without producing *p*, and therefore (unless counteracted somehow) cannot be a condition of *p*. A, then, stands out as the one thing that is present whenever *p* is present, and absent whenever *p* is absent.

Stated in this abstract way, the Double Method may seem very elaborate and difficult; yet, in fact, its use may be very simple. Tyndall, to prove that dispersed light in the air is due to motes, showed by a number of cases (1) that any gas containing motes is luminous; (2) that air in which the motes had been destroyed by heat, and any gas so prepared as to exclude motes, are not luminous. All the instances are of gases, and the result is: motes—luminosity; no motes—no luminosity. Darwin, to show that cross-fertilisation is favourable to flowers, placed a net about 100 flower-heads, and left 100 others of the same varieties exposed to the bees: the former bore no seed, the latter nearly 3,000. We must assume that, in Darwin's judgment, the net did not screen the flowers from light and heat sufficiently to affect the result.

There are instructive applications of this Double Method in Wallace's *Darwinism*. In chap. viii., on *Colour in Animals*, he observes, that the usefulness of their coloration to animals is shown by the fact that, "as a rule, colour and marking are constant in each species of wild animal, while, in almost every domesticated animal, there arises great variability. We see this in our horses and cattle, our dogs and cats, our pigeons and poultry. Now the essential difference between the conditions of life of domesticated and wild animals is, that the former are protected by man, while the latter have to protect themselves." Wild animals protect themselves by acquiring qualities adapted to their mode of life; and coloration is a very important one, its chief, though not its only use, being concealment. Hence a useful coloration having been established in any species, individuals that occasionally may vary from it, will generally, perish; whilst, among domestic animals, variation of colour or marking is subject to no check except the taste of owners. We have, then, two lists of instances; first, innumerable species of wild animals in which the coloration is constant and which depend upon their own qualities for existence; secondly, several species of domestic animals in which the coloration is *not* constant, and which do *not* depend upon their own qualities for existence. In the former list two circumstances are present together (under all sorts of conditions); in the latter they are absent together. The argument may be further strengthened by adding a third list, parallel to the first, comprising domestic animals in which coloration is approximately constant, but where (as we know) it is made a condition of existence by owners, who only breed from those specimens that come up to a certain standard of coloration.

Wallace goes on to discuss the colouring of arctic animals. In the arctic regions, he says, some animals are wholly white all the year round, such as the polar bear, the American polar hare, the snowy owl and the Greenland falcon: these live amidst almost perpetual snow. Others, that live where the snow melts in summer, only turn white in winter, such as the arctic hare, the arctic fox, the ermine and the ptarmigan. In all these cases the white colouring is useful, concealing the herbivores from their enemies, and also the carnivores in approaching their prey; this usefulness, therefore, is a condition of the white colouring. Two other explanations have, however, been suggested: first, that the prevalent white of the arctic regions directly colours the animals, either by some photographic or chemical action on the skin, or by a reflex action through vision (as in the chameleon); secondly, that a white skin checks radiation and keeps the animals warm. But there are some exceptions to the rule of white colouring in arctic animals which refute these hypotheses, and confirm the author's. The sable remains brown throughout the winter; but it frequents trees, with whose bark its colour assimilates. The musk-sheep is brown and conspicuous; but it is gregarious, and its safety depends upon its ability to recognise its kind and keep with the herd. The raven is always black; but it fears no enemy and feeds on carrion, and therefore does not need concealment for either defence or attack. The colour of the sable, then, though not white, serves for concealment; the colour of the musk-sheep serves a purpose more important than concealment; the raven needs no concealment. There are thus two sets of instances:—in one set the animals are white (*a*) all the year, (*b*) in winter; and white conceals them (*a*) all the year, (*b*) in winter; in the other set, the animals are *not* white, and to them either whiteness would *not* give concealment, or concealment would *not* be advantageous. And this second list refutes the rival hypotheses: for the sable, the musk-sheep and the raven are as much exposed to the glare of the snow, and to the cold, as the other animals are.

§ 3. The Canon of Difference.

If an instance in which a phenomenon occurs, and an instance in which it does not occur, have every other circumstance in common save one, that one (whether consequent or antecedent) occurring only in the former; the circumstance in which alone the two instances differ is the effect, or the cause, or an indispensable condition of the phenomenon.

This follows from Props. I (*a*) and (*b*), in chapter xv. § 7. To prove that A is a condition of *p*, let two instances, such as the Canon requires, be represented thus:

Then A is the cause or a condition of p. For, in the first instance, A being introduced (without further change), p arises (Prop. I. (a)); and, in the second instance, A having been removed (without other change), p disappears (Prop. I. (b)). Similarly we may prove, by the same instances, that p is the effect of A.

The order of the phenomena and the immediacy of their connection is a matter for observation, aided by whatever instruments and methods of inspection and measurement may be available.

As to the invariability of the connection, it may of course be tested by collecting more instances or making more experiments; but it has been maintained, that a single perfect experiment according to this method is sufficient to prove causation, and therefore implies invariability (since causation is uniform), though no other instances should ever be obtainable; because it establishes once for all the unconditionality of the connection

Now, formally this is true; but in any actual investigation how shall we decide what is a satisfactory or perfect experiment? Such an experiment requires that in the negative instance

BC shall be the least assemblage of conditions necessary to co-operate with A in producing p; and that it is so cannot be ascertained without either general prior knowledge of the nature of the case or special experiments for the purpose. So that invariability will not really be inferred from a single experiment; besides that every prudent inquirer repeats his experiments, if only to guard against his own liability to error.

The supposed plurality of causes does not affect the method of Difference. In the above symbolic case, A is clearly *one* cause (or condition) of p, whatever other causes may be possible; whereas with the Single Method of Agreement, it remained doubtful (admitting a plurality of causes) whether A, in spite of being always present with p, was ever a cause or condition of it.

This method of Difference without our being distinctly aware of it, is oftener than any other the basis of ordinary judgments. That the sun gives light and heat, that food nourishes and fire burns, that a stone breaks a window or kills a bird, that the turning of a tap permits or checks the flow of water or of gas, and thousands of other propositions are known to be true by rough but often emphatic applications of this method in common experience.

The method of Difference may be applied either (1) by observation, on finding two instances (distinct assemblages of conditions) differing only in one phenomenon together with its antecedent or consequent; or (2) by experiment, and then, either (a) by preparing two instances that may be compared side by side, or (b) by taking certain conditions, and then introducing (or subtracting) some agent, supposed to be the cause, to see what happens: in the latter case the "two instances" are the same assemblage of conditions considered before and, again, after, the introduction of the agent. As an example of (a) there is an experiment to show that radium gives off heat: take two glass tubes, in one put some chloride of radium, in both thermometers, and close them with cotton-wool. Soon the thermometer in the tube along with radium reads 54° F. higher than the other one. The tube without the radium, whose temperature remains unaltered, is called the "control" experiment. Most experiments are of the type (b); and since the Canon, which describes two co-existing instances, does not readily apply to this type, an alternative version may be offered: *Any agent whose introduction into known circumstances (without further change) is immediately followed by a definite phenomenon is a condition of the occurrence of that phenomenon.*

The words *into known circumstances* are necessary to emphasise what is required by this Method, namely, that the two instances differ in only one thing; for this cannot be ascertained unless all the other conditions are known; and this further implies that they have been prepared. It is, therefore, not true (as Sigwart asserts) that this method determines only one condition of a phenomenon, and that it is then necessary to inquire into the other conditions. If they were not known they must be investigated; but then the experiment would not have been made upon this method. Practically, experiments have to be made in all degrees of imperfection, and the less perfect they are, that is, the less the circumstances are known beforehand, the more remains to be done. A common imperfection is delay, or the occurrence of a latent period between the introduction of an agent and the manifestation of its effects; it cannot then be the unconditional cause; though it may be an indispensable remote condition of whatever change occurs. If, feeling out of sorts, you take a drug and some time afterwards feel better, it is not clear on this ground alone that the drug was the cause of recovery, for other curative processes may have been active meanwhile—food, or sleep, or exercise.

Any book of Physics or of Chemistry will furnish scores of examples of the method of Difference: such as Galileo's experiment to show that air has weight, by first weighing a vessel filled with ordinary air, and then filling it with condensed air and weighing it again; when the increased weight can only be due to the greater quantity of air contained. The melting-point of solids is determined by heating them until they do melt (as silver at 1000° C., gold at 1250°, platinum at 2000°); for the only difference between bodies at the time of melting and just before is the addition of so much heat. Similarly with the boiling point of liquids. That the transmission of sound depends upon the continuity of an elastic ponderable medium, is proved by letting a clock strike in a vacuum (under a glass from which the air has been withdrawn by an air pump), and standing upon a non-elastic pedestal: when the clock be seen to strike, but makes only such a faint sound as may be due to the imperfections of the vacuum and the pedestal.

The experiments by which the chemical analysis or synthesis of various forms of matter is demonstrated are simple or compound applications of this method of Difference, together with the quantitative mark of causation (that cause and effect are equal); since the bodies resulting from an analysis are equal in weight to the body analysed, and the body resulting from a synthesis is equal in weight to the bodies synthesised. That an electric current resolves water into oxygen and hydrogen may be proved by inserting the poles of a galvanic battery in a vessel of water; when this one change is followed by another, the rise of bubbles from each pole and the very gradual decrease of the water. If the bubbles are caught in receivers placed over them, it can be shown that the joint weight of the two bodies of gas thus formed is equal to the weight of the water that has disappeared; and that the gases are respectively oxygen and hydrogen may then be shown by proving that they have the properties of those gases according to further experiments by the method of Difference; as (*e.g.*) that one of them is oxygen because it supports combustion, *etc*.

When water was first decomposed by the electric current, there appeared not only oxygen and hydrogen, but also an acid and an alkali. These products were afterwards traced to impurities of the water and of the operator's hands. Mill observes that in any experiment the effect, or part of it, may be due, not to the supposed agent, but to the means employed in introducing it. We should know not only the other conditions of an experiment, but that the agent or change introduced is nothing else than what it is supposed to be.

In the more complex sciences the method of Difference is less easily applicable, because of the greater difficulty of being sure that only one circumstance at a time has altered; still, it is frequently used. Thus, if by dividing a certain nerve certain muscles are paralysed, it is shown that normally that nerve controls those muscles. That the sense of smell in flies and cockroaches is connected with the antennae has been shown by cutting them off: whereupon the insects can no longer find carrion. In his work on *Earthworms*, Darwin shows that, though sensitive to mechanical tremors, they are deaf (or, at least, not sensitive to sonorous vibrations transmitted through the air), by the following experiment. He placed a pot containing a worm that had come to the surface, as usual at night, upon a table, whilst close by a piano was violently played; but the worm took no notice of the noise. He then placed the pot upon the piano, whilst it was being played, when the worm, probably feeling mechanical vibrations, hastily slid back into its burrow.

When, instead of altering one circumstance in an instance (which we have done our best not otherwise to disturb) and then watching what follows, we try to find two ready-made instances of a phenomenon, which only differ in one other circumstance, it is, of course, still more difficult to be sure that there is only one other circumstance in which they differ. It may be worth while, however, to look for such instances. Thus, that the temperature of ocean currents influences the climate of the shores they wash, seems to be shown by the fact that the average temperature of Newfoundland is lower than that of the Norwegian coast some 15° farther north. Both regions have great continents at their back; and as the mountains of Norway are higher and capped with perennial snow, we might expect a colder climate there: but the shore of Norway is visited by the Gulf Stream, whilst the shore of Newfoundland is traversed by a cold current from Greenland. Again, when in 1841 the railway from Rouen to Paris was being built, gangs of English and gangs of French workmen were employed upon it, and the English got through about one-third more work per man than the French. It was suspected that this difference was due to one other difference, namely, that the English fed better, preferring beef to thin soup. Now, logically, it might have been objected that the evidence was unsatisfactory, seeing that the men differed in other things besides diet—in 'race' (say), which explains so much and so easily. But the Frenchmen, having been induced to try the same diet as the English, were, in a few days, able to do as much work: so that the "two instances" were better than they looked. It often happens that evidence, though logically questionable, is good when used by experts, whose familiarity with the subject makes it good.

§ 4. The Canon Of Concomitant Variations.

Whatever phenomenon varies in any manner whenever another phenomenon (consequent or antecedent) varies in some particular manner is either the cause or effect of that phenomenon.

This is not an entirely fresh method, but may be regarded as a special case either of Agreement or of Difference, to prove the cause or effect, not of a phenomenon as a whole, but of some increment of it (positive or negative). There are certain forces, such as gravitation, heat, friction, that can never be eliminated altogether, and therefore can only be studied in their degrees. To such phenomena the method of Difference cannot be applied, because there are no negative instances. But we may obtain negative instances of a given quantity of such a phenomenon (say, heat), and may apply the method of Difference to that quantity. Thus, if the heat of a body increases 10 degrees, from 60 to 70, the former temperature of 60 was a negative instance in respect of those 10 degrees; and if only one other circumstance (say, friction) has altered at the same time, that circumstance (if an antecedent) is the cause. Accordingly, if in the above Canon we insert, after 'particular manner,' "" it is a statement of the method of Difference as applicable to the increment of a phenomenon, instead of to the phenomenon as a whole; and we may then omit the last clause—"." For these words are inserted to provide for the case of co-effects of a common cause (such as the flash and report of a gun); but if no other change (such as the discharge of a gun) has concurred with the variations of two phenomena, there cannot have been a common cause, and they are therefore cause and effect.

If, on the other hand, we omit the clause "" the Canon is a statement of the method of Agreement as applicable to the increment of a phenomenon instead of to the phenomenon as a whole; and it is then subject to the imperfections of that method: that is to say, it leaves open the possibilities, that an inquirer may overlook a plurality of causes; or may mistake a connection of two phenomena, which (like the flash and report of a gun) are co-effects of a common cause, for a direct relation of cause and effect.

It may occur to the reader that we ought also to distinguish Qualitative and Quantitative Variations as two orders of phenomena to which the present method is applicable. But, in fact, Qualitative Variations may be adequately dealt with by the foregoing methods of Agreement, Double Agreement, and Difference; because a change of quality or property entirely gets rid of the former phase of that quality, or substitutes one for another; as when the ptarmigan changes from brown to white in winter, or as when a stag grows and sheds its antlers with the course of the seasons. The peculiar use of the method of Variations, however, is to formulate the conditions of proof in respect of those causes or effects which cannot be entirely got rid of, but can be obtained only in greater or less amount; and such phenomena are or course, quantitative.

Even when there are two parallel series of phenomena the one quantitative and the other qualitative—like the rate of air-vibration and the pitch of sound, or the rate of ether-vibration and the colour-series of the spectrum—the method of Variations is not applicable. For (1) two such series cannot be said to vary together, since the qualitative variations are heterogeneous: 512: 576 is a definite ratio; but the corresponding notes, C, D, in the treble clef, present only a difference. Hence (2) the correspondence of each note with each number is a distinct fact. Each octave even is a distinct fact; there is a difference between C 64 and C 128 that could never have been anticipated without the appropriate experience. There is, therefore, no such law of these parallel series as there is for temperature and change of volume (say) in mercury. Similar remarks apply to the physical and sensitive light-series.

We may illustrate the two cases of the method thus (putting a dash against any letter, A' or p', to signify an increase or decrease of the phenomenon the letter stands for): Agreement in Variations (other changes being admissible)—

$$A \mid B\,C\,q\,r \mid A'\,D\,E\,s\,t \mid A''\,F\,G\,u\,v$$

$$p \mid \quad\quad\quad \mid p' \quad\quad\quad \mid p''$$

Here the accompanying phenomena (*B C q r, D E s t, F G u v*) change from time to time, and the one thing in which the instances agree throughout is that any increase of A (A' or A") is followed or accompanied by an increase of *p (p' or p")*: whence it is argued that A is the cause of *p*, according to Prop. III. (*a*) (ch. xv. § 7). Still, it is supposable that, in the second instance, D or E may be the cause of the increment of *p*; and that, in the third instance, F or G may be its cause: though the probability of such vicarious causation decreases rapidly with the increase of instances in which A and *p* vary together. And, since an actual investigation of this type must rely on observation, it is further possible that some undiscovered cause, X, is the real determinant of both A and *p* and of their concomitant variations.

Professor Ferri, in his *Criminal Sociology*, observes: "I have shown that in France there is a manifest correspondence of increase and decrease between the number of homicides, assaults and malicious wounding, and the more or less abundant vintage, especially in the years of extraordinary variations, whether of failure of the vintage (1853-5, 1859, 1867, 1873, 1878-80), attended by a remarkable diminution of crime (assaults and wounding), or of abundant vintages (1850, 1856-8, 1862-3, 1865, 1868, 1874-5), attended by an increase of crime" (p. 117, Eng. trans.). And earlier he had remarked that such crimes also "in their oscillations from month to month display a characteristic increase during the vintage periods, from June to December, notwithstanding the constant diminution of other offences" (p. 77). This is necessarily an appeal to the canon of Concomitant Variations, because France is never without her annual vintage, nor yet without her annual statistics of crime. Still, it is an argument whose cogency is only that of Agreement, showing that probably the abuse of the vintage is a cause of crimes of violence, but leaving open the supposition, that some other circumstance or circumstances, arising or varying from year to year, may determine the increase or decrease of crime; or that there is some unconsidered agent which affects both the vintage and crimes of violence. French sunshine, it might be urged, whilst it matures the generous grape, also excites a morbid fermentation in the human mind.

Difference in Variations may be symbolically represented thus (no other change having concurred):

$$A \mid B \mid A'\,B \mid A''\,B$$

$$q \mid \quad \mid p'\,q \mid p''\,q$$

Here the accompanying phenomena are always the same B/q; and the only point in which the successive instances differ is in the increments of A (A', A") followed by corresponding increments of *p (p', p")*: hence the increment of A is the cause of the increment of *p*.

For examples of the application of this method, the reader should refer to some work of exact science. He will find in Deschanel's *Natural Philosophy*, c. 32, an account of some experiments by which the connection between heat and mechanical work has been established. It is there shown that "whenever work is performed by the agency of heat" , "an amount of heat disappears equivalent to the work performed; and whenever mechanical work is spent in generating heat" , "the heat generated is equivalent to the work thus spent." And an experiment of Joule's is described, which consisted in fixing a rod with paddles in a vessel of water, and making it revolve and agitate the water by means of a string wound round the rod, passed over a pulley and attached to a weight that was allowed to fall. The descent of the weight was measured by a graduated rule, and the rise of the water's temperature by a thermometer. "It was found that the heat communicated to the water by the agitation amounted to one pound-degree Fahrenheit for every 772 foot-pounds of work" expended by the falling weight. As no other material change seems to take place during such an experiment, it shows that the progressive expenditure of mechanical energy is the cause of the progressive heating of the water.

The thermometer itself illustrates this method. It has been found that the application of heat to mercury expands it according to a law; and hence the volume of the mercury, measured by a graduated index, is used to indicate the temperature of the air, water, animal body, *etc.*, in which the thermometer is immersed, or with which it is brought into contact. In such cases, if no other change has taken place, the heat of the air, water, or body is the cause of the rise of the mercury in its tube. If some other substance (say spirit) be substituted for mercury in constructing a thermometer, it serves the same purpose, provided the index be graduated according to the law of the expansion of that substance by heat, as experimentally determined.

Instances of phenomena that do not vary together indicate the exclusion of a supposed cause (by Prop. III (*c*)). The stature of the human race has been supposed to depend on temperature; but there is no correspondence. The "not varying together," however, must not be confused with "varying inversely," which when regular indicates a true concomitance. It is often a matter of convenience whether we regard concomitant phenomena as varying directly or inversely. It is usual to say—'the greater the friction the less the speed'; but it is really more intelligible to say—'the greater the friction the more rapidly molar is converted into molecular motion.'

The Graphic Method exhibits Concomitant Variations to the eye, and is extensively used in physical and statistical inquiries. Along a horizontal line (the abscissa) is measured one of the conditions (or agents) with which the inquiry is concerned, called the Variable; and along perpendiculars (ordinates) is measured some phenomenon to be compared with it, called the Variant.

Thus, the expansion of a liquid by heat may be represented by measuring degrees of temperature along the horizontal, and the expansion of a column of the liquids in units of length along the perpendicular.

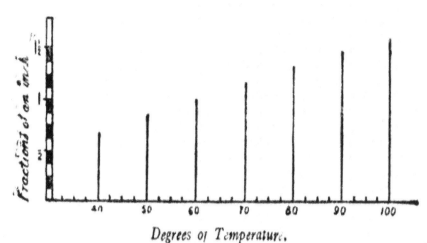

Fig. 9.

In the next diagram (Fig. 10), reduced from one given by Mr. C.H. Denyer in an article on the Price of Tea (*Economic Journal*, No. 9), the condition measured horizontally is Time; and, vertically, three variants are measured simultaneously, so that their relations to one another from time to time may be seen at a glance. From this it is evident that, as the duty on tea falls, the price of tea falls, whilst the consumption of tea rises; and, in spite of some irregularity of correspondence in the courses of the three phenomena, their general causal connection can hardly be mistaken. However, the causal connection may also be inferred by general reasoning; the statistical Induction can be confirmed by a Deduction; thus illustrating the combined method of proof to be discussed in the next chapter. Without such confirmation the proof by Concomitant Variations would not be complete; because, from the complexity of the circumstances, social statistics can only yield evidence according to the method of Agreement in Variations. For, besides the agents that are measured, there may always be some other important influence at work. During the last fifty years, for example, crime has decreased whilst education has increased: true, but at the same time wages have risen and many other things have happened.

Diagram showing (1)— · — · the average Price of Tea (in bond), but with duty added per lb.; (2)· · · · · the rate of Duty; (3)---------- the consumption per head, from 1809 to 1889.

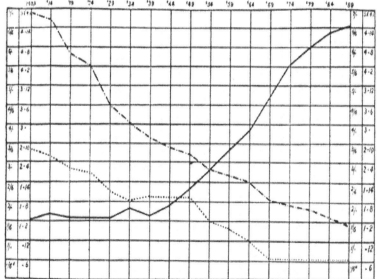

Fig. 10.

One horizontal space = 5 years. One vertical space = 6 pence, or 6 ounces.

It will be noticed that in the diagram the three lines, especially those of Price and Consumption (which may be considered *natural* resultants, in contrast with the arbitrary fixation of a Tax), do not depart widely from regular curves; and accordingly, assuming the causes at work to vary continuously during the intervals between points of measurement, curves may be substituted. In fact, a curve often represents the course of a phenomenon more truthfully than can be done by a line that zigzags along the exact measurements; because it is less influenced by temporary and extraordinary causes that may obscure the operation of those that are being investigated. On the other hand, the abrupt deviations of a punctilious zigzag may have their own logical value, as will appear in the next section.

In working with the Method of Variations one must allow for the occurrence in a series of 'critical points,' at which sudden and sometimes heterogeneous changes may take place. Every substance exists at different temperatures in three states, gaseous, liquid, solid; and when the change takes place, from one state to another, the series of variations is broken. Water, *e.g.*, follows the general law that cooling is accompanied by decrease of volume between 212° and 39° F.: but above 212°, undergoes a sudden expansion in becoming a gas;

and below 39° begins to expand, until at 32° the expansion is considerable on its becoming solid. This illustrates a common experience that concomitant variations are most regular in the 'median range,' and are apt to become irregular at the extremities of the series, where new conditions begin to operate.

The Canon of Variations, again, deals not with sudden irruptions of a cause, force or agent, but with some increase or decrease of an agent already present, and a corresponding increase or decrease of some other phenomenon—say an increase of tax and a rise of price. But there are cases in which the energy of a cause is not immediately discharged and dissipated. Whilst a tax of $6d.$ per lb. on tea raises the price per lb. by about $6d.$, however long it lasts, the continuous application of friction to a body may gradually raise its temperature to the point of combustion; because heat is received faster than it is radiated, and therefore accumulates. Such cases are treated by Mill under the title of 'progressive effects' (*Logic*: B. III., c. 15): he gives as an example of it the acceleration of falling bodies. The storage of effects is a fact of the utmost importance in all departments of nature, and is especially interesting in Biology and Sociology, where it is met with as heredity, experience, tradition. Evolution of species of plants and animals would (so far as we know) be impossible, if the changes (however caused) that adapt some individuals better than others to the conditions of life were not inherited by, and accumulated in, their posterity. The eyes in the peacock's tail are supposed to have reached their present perfection gradually, through various stages that may be illustrated by the ocelli in the wings of the Argus pheasant and other genera of *Phasianidæ*. Similarly the progress of societies would be impossible without tradition, whereby the improvements made in any generation may be passed on to the next, and the experience of mankind may be gradually accumulated in various forms of culture. The earliest remains of culture are flint implements and weapons; in which we can trace the effect of tradition in the lives of our remote forefathers, as they slowly through thousands of years learnt to improve the chipping of flints, until the first rudely shaped lumps gave place to works of unmistakable design, and these to the beautiful weapons contemporary with the Bronze Age.

The Method of Gradations, the arranging of any phenomena to be studied in series, according to the degree in which some character is exhibited, is, perhaps, the most definite device in the Art of Discovery. (Bain: *Induction*, c. 6, and App. II.) If the causes are unknown it is likely to suggest hypotheses: and if the causes are partly known, variation in the character of the series is likely to indicate a corresponding variation of the conditions.

§ 5. The Canon Of Residues.

Subduct from any phenomenon such part as previous inductions have shown to be the effect of certain antecedents, and the residue of the phenomenon is the effect of the remaining antecedents.

The phenomenon is here assumed to be an effect: a similar Canon may be framed for residuary causes.

This also is not a fresh method, but a special case of the method of Difference. For if we suppose the phenomenon to be $p\ q\ r$, and the antecedent to be A B C, and that we already know B and C to have (either severally or together) the consequents $q\ r$, in which their efficacy is exhausted; we may regard

| (
(|

as an instance of the absence of p obtained deductively from the whole phenomenon

| | (
| (|

by our knowledge of the laws of B and C; so that

| | (
| (|

is an instance of the presence of p, differing otherwise from

| (
(|

in nothing except that A is also present. By the Canon of Difference, therefore A is the cause of p. Or, again, when phenomena thus treated are strictly quantitative, the method may be based on Prop. III. (*b*), ch. xv. § 7.

Of course, if A can be obtained apart from B C and directly experimented with so as to produce p, so much the better; and this may often be done; but the special value of the method of Residues appears, when some complex phenomenon has been for the most part accounted for by known causes, whilst there remains some excess, or shortcoming, or deviation from the result which those causes alone would lead us to expect, and this residuary fact has to be explained in relation to the whole. Here the negative instance is constituted by deduction, showing what would happen but for the interference of some unknown cause which is to be investigated; and this prominence of the deductive process has led some writers to class the method as deductive. But we have seen that all the Canons involve deduction; and, considering how much in every experiment is assumed as already known (what circumstances are 'material,' and when conditions may be called 'the same'), the wonder is that no one has insisted upon regarding every method as concerned with residues. In fact, as scientific explanation progresses, the phenomena that may be considered as residuary become more numerous and the importance of this method increases.

Examples: The recorded dates of ancient eclipses having been found to differ from those assigned by calculation, it appears that the average length of a day has in the meanwhile increased. This is a residuary phenomenon not accounted for by the causes formerly recognised as determining the rotation of the earth on its axis; and it may be explained by the consideration that the friction of the tides reduces the rate of the earth's rotation, and thereby lengthens the day. Astronomy abounds in examples of the method of Residues, of which the discovery of Neptune is the most famous.

Capillarity seems to be a striking exception to the principle that water (or any liquid) 'finds its level,' that being the condition of equilibrium; yet capillarity proves to be only a refined case of equilibrium when account is taken of the forces of adhesion exerted by different kinds of bodies in contact.

"Many of the new elements of Chemistry," says Herschel, "have been detected in the investigation of residual phenomena." Thus, Lord Rayleigh and Sir W. Ramsay found that nitrogen from the atmosphere was slightly heavier than nitrogen got from chemical sources; and, seeking the cause of this difference, discovered argon.

The Economist shows that when a country imports goods the chief means of paying for them is to export other goods. If this were all, imports and exports would be of equal value: yet the United Kingdom imports about £400,000,000 annually, and exports about £300,000,000. Here, then, is a residuary phenomenon of £100,000,000 to be accounted for. But foreign countries owe us about £50,000,000 for the use of shipping, and £70,000,000 as interest on the capital we have lent them, and £15,000,000 in commissions upon business transacted for them. These sums added together amount to £135,000,000; and that is £35,000,000 too much. Thus another residuary phenomenon emerges; for whilst foreigners seem to owe us £435,000,000 they only send us £400,000,000 of imports. These £35,000,000 are accounted for by the annual investment of our capital abroad, in return for which no immediate payment is due; and, these being omitted, exports and imports balance. Since this was written the figures of our foreign trade have greatly risen; but the character of the explanation remains the same.

When, in pursuing the method of Variations, the phenomena compared do not always correspond in their fluctuations, the irregular movements of that phenomenon which we regard as the effect may often be explained by treating them as residuary phenomena, and then seeking for exceptional causes, whose temporary interference has obscured the influence of the general cause. Thus, returning to the diagram of the Price of Tea in § 4, it is clear that generally the price falls as the duty falls; but in Mr. Denyer's more minutely wrought diagram, from which this is reduced, it may be seen that in 1840 the price of tea rose from 3s. 9d. to 4s. 9d. without any increase of duty. This, however, is readily explained by the Chinese War of that year, which checked the supply. Again, from 1869 to 1889 the duty was constant, whilst the price of tea fell as much as 8d. per lb.; but this residuary phenomenon is explained by the prodigiously increased production of tea during that period in India and Ceylon.

The above examples of the method of Residues are all quantitative; but the method is often employed where exact estimates are unobtainable. Thus Darwin, having found certain modifications of animals in form, coloration and habits, that were not clearly derivable from their struggle for existence in relation to other species or to external conditions, suggested that they were due to Sexual Selection.

The 'vestiges' and 'survivals' so common in Biology and Sociology are residuary phenomena. It is a general inference from the doctrine of Natural Selection that every organ of a plant, animal, or society is in some way useful to it. There occur, however, organs that have at present no assignable utility, are at least wasteful, and sometimes even injurious. And the explanation is that formerly they were useful; but that, their uses having lapsed, they are now retained by the force of heredity or tradition. Either they are not injurious enough to be eliminated by natural selection; or they are correlated with other organs, whose utility outweighs their disutility.

CHAPTER XVII
COMBINATION OF INDUCTION WITH DEDUCTION

§ 1. We have now reviewed Mill's five Canons of Inductive Proof. At bottom, as he observes, there are only two, namely, Agreement and Difference: since the Double Method, Variations and Residues are only special forms of the other two. Indeed, in their function of *proof*, they are all reducible to one, namely, Difference; for the cogency of the method of Agreement (as distinguished from a simple enumeration of instances agreeing in the coincidence of a supposed cause and its effect), depends upon the omission, in one instance after another, of all other circumstances; which omission is a point of difference.

The Canons are an analysis of the conditions of proving directly (where possible), by means of observation or experiment, any proposition that predicates causation. But if we say 'by means of observation or experiment,' it is not to be understood that these are the only means and that nothing else is involved; for it has been shown that the Law of Causation is itself an indispensable foundation of the evidence. In fact Inductive Logic may be considered as having a purely formal character. It consists (1) in a statement of the Law of Cause and Effect; (2) in certain immediate inferences from this Law, expanded into the Canons; (3) in the syllogistic application of the Canons to special predications of causation by means of minor premises, showing that certain instances satisfy the Canons.

At the risk of some pedantry, we may exhibit the process as follows (*cf.* Prof. Ray's *Logic*: Appendix D):
Whatever relation of events has certain marks is a case of causation;
The relation A: p has some or all of these marks (as shown by observation and by the conformity of instances to such or such a Canon):
Therefore, the relation A: p is a case of causation. Now, the parenthesis, "as shown by the conformity, *etc.*," is an adscititious member of an Epicheirema, which may be stated, as a Prosyllogism, thus:
If an instance, etc. (Canon of Difference);
The instances

/ ! (　! (

! (,′　(/

are of the kind required:
Therefore, A, present where p occurs and absent where it does not occur, is an indispensable antecedent of p.

Such is the bare Logic of Induction: so that, strictly speaking, observation or experiment is no part of the logic, but a means of applying the logic to actual, that is, not merely symbolical, propositions. The Formal Logic of Induction is essentially deductive; and it has been much questioned whether any transition from the formal to the material conditions of proof is possible. As long as we are content to illustrate the Canons with symbols, such as A and *p*, all goes well; but can we in any actual investigation show that the relevant facts or 'instances' correspond with those symbols?

In the first place, as Dr. Venn shows, natural phenomena want the distinctness and capability of isolation that belong to symbols. Secondly, the observing whether instances conform to a Canon, must always be subject at last to the limits of our faculties. How can we ascertain exact equality, immediate sequence? The Canon of Difference, in its experimental application, is usually considered the most cogent sort of proof: yet when can the two sequent instances, before and after the introduction of a certain agent, be said to differ in nothing else? Are not earth and stars always changing position; is not every molecule in the room and apparatus always oscillating? It is true that our senses are now aided by elaborate instruments; but the construction of these depends on scientific theories, which again depend on experiments.

It is right to touch upon this well-known sceptical topic; but to insist much upon it is not a sign of good sense. The works of Herschel, Whewell, and Jevons should be consulted for the various methods of correcting observations, by repeating them, averaging them, verifying one experimental process by another, always refining the methods of exact measurement, multiplying the opportunities of error (that if any exist it may at last show itself), and by other devices of what may be called Material Logic or Methodology. But only direct experience and personal manipulation of scientific processes, can give a just sense of their effectiveness; and to stand by, suggesting academic doubts, is easier and more amusing.

§ 2. Still, it is not so much in laws based upon direct observation or experiment, that the material validity of scientific reasoning appears, as in the cumulative evidence that arises from the co-ordination of laws within each science, and the growing harmony and coherence of all sciences. This requires a more elaborate combination of deduction with observation and experiment. During the last three hundred years many departments of science have been reduced under principles of the greatest generality, such as the Conservation of Energy, the Law of Gravitation, the Undulatory theory of Light, the Law of combining Equivalents, and the Theory of Natural Selection; connecting and explaining the less general laws, which, again, are said to connect and explain the facts. Meanwhile, those sciences that were the first to make progress have helped to develop others which, like Biology and Sociology, present greater difficulties; and it becomes more and more apparent that the distinctions drawn among sciences are entirely for the convenience of study, and that all sciences tend to merge in one universal Science of Nature. Now, this process of the 'unification of knowledge' is almost another name for deduction; but at the same time it depends for its reality and solidity upon a constant reference to observation and experiment. Only a very inadequate notion of it can be given in the ensuing chapters.

We saw in chap. xiv. § 6, that when two or more agents or forces combine to produce a phenomenon, their effects are intermixed in it, and this in one of two ways according to their nature. In chemical action and in vegetable and animal life, the causal agents concerned are blended in their results in such a way that most of the qualities which they exhibited severally are lost, whilst new qualities appear instead. Thus chlorine (a greenish-yellow gas) and sodium (a metal) unite to form common salt $NaCl$; which is quite unlike either of them: a man eats bread, and it becomes muscle, nerve and bone. In such cases we cannot trace the qualities of the causal agents in the qualities of the effects; given such causes, we can prove experimentally, according to the canons of induction, that they have such effects; but we may not be able in any new case to calculate what the effects will be.

On the other hand, in Astronomy and Physics, the causes treated of are mechanical; at least, it is the aim of Physics to attain to a mechanical conception of phenomena; so that, in every new combination of forces, the intermixed effect, or resultant, may be calculated beforehand; provided that the forces concerned admit of being quantitatively estimated, and that the conditions of their combination are not so complex as to baffle the powers of mathematicians. In such cases, when direct observation or experiment is insufficient to resolve an effect into the laws of its conditions, the general method is to calculate what may be expected from a combination of its conditions, as either known or hypothetically assumed, and to compare this anticipation with the actual phenomenon.

§ 3. This is what Mill calls the Direct Deductive Method; or, the Physical Method, because it is so much relied on in treating of Light, Heat, Sound, *etc.*; it is also the method of Astronomy and much used in Economics: Deduction leads the way, and its results are tested inductively by experiments or observations. Given any complex mechanical phenomenon, the inquirer considers—(1) what laws already ascertained seem likely to apply to it (in default of known laws, hypotheses are substituted: *cf.* chap. xviii.); he then—(2) computes the effect that will follow from these laws in circumstances similar to the case before him; and (3) he verifies his conclusion by comparing it with the actual phenomenon.

A simple example of this method is the explanation of the rise of water in the 'common pump.' We know three laws applicable to this case: (*a*) that the atmosphere weighs upon the water outside the pump with a pressure of 15 lb. to the square inch; (*b*) that a liquid (and therefore the water) transmits pressure equally in all directions (upwards as well as downwards and sideways); and (*c*) that pressure upon a body in any direction, if not counteracted by an opposite pressure, produces motion. Hence, when the rise of the piston of the pump removes the pressure upon the water within the cylinder, tending to produce a vacuum there, this water is pushed up by the pressure of the air upon the water outside the cylinder, and follows the rising piston, until the column of water inside the cylinder exerts a pressure equal to that of the atmosphere upon an equal area. So much for the computation; does it correspond with the fact? It is found that at the sea level water can be pumped to the height of 33 ft; and that such a column of water has a pressure of 15 lb. to the square inch. We may show further that, at the sea level, spirits of wine may be pumped higher according to its less specific gravity; and that if we attempt to pump water at successive altitudes above the sea level, we can only raise it to less and less heights, corresponding with the lessened atmospheric pressure at those altitudes, where the column of air producing the pressure is shorter. Finally, if we try to work a pump, having first produced a vacuum over the water outside the cylinder, we shall find that the water inside will not rise at all; the piston can be raised, but the water does not follow it. The verification thus shows that the computed effect corresponds with the phenomenon to be explained; that the result does not depend upon the nature of water only, but is true (allowing for differences of specific gravity) of other liquids; that if the pressure of the outside air is diminished, the height of pumping is so too (canon of Variations); and that if that pressure is entirely removed, pumping becomes impossible (canon of Difference).

Any text-book of Astronomy or Physics furnishes numerous illustrations of the deductive method. Take, for example, the first chapter of Deschanel's *Optics*, where are given three methods of determining the velocity of Light. This was first deduced from observation of Jupiter's satellites. The one nearest the planet passes behind it, or into its shadow, and is eclipsed, at intervals of about 42½ hours. But it can be shown that, when Jupiter and the Earth are nearest together on the same side of the Sun, an eclipse of this satellite is visible from the earth 16 min. 26.6 sec. earlier than when Jupiter and the earth are furthest apart on opposite sides of the Sun: 16 min. 26.6 sec, then, is the time in which light traverses the diameter of the Earth's orbit. Therefore, supposing the Earth's distance from the Sun to be 92 millions of miles, light travels about 186,000 miles a second. Another deduction, agreeing with this, starts from the fact of aberration, or the displacement of the apparent from the actual position of the stars in the direction of the earth's motion. Aberration depends partly on the velocity of light, partly on the velocity of the Earth; and the latter being known, the former can be computed. Now, these two deductive arguments, verifying each other, have also been verified experimentally. Foucault's experiment to measure the velocity of light is too elaborate to be described here: a full account of it will be found in the treatise above cited, § 687.

When the phenomena to be explained are of such a character, so vast in extent, power or duration, that it is impossible, in the actual circumstances of the case, to frame experiments in order to verify a deductive explanation, it may still be possible to reproduce a similar phenomenon upon a smaller scale. Thus Monge's explanation of mirage by the great heat of the desert sand, which makes the lowest stratum of air less dense than those above it, so that rays of light from distant objects are refracted in descending, until they are actually turned upwards again to the eye of the beholders, giving him inverted images of the objects as if they were reflected in water, is manifestly incapable of being verified by experiment in the natural conditions of the phenomenon. But by heating the bottom of "a sheet-iron box, with its ends cut away," the rarefied air at the bottom of the box may sometimes be made to yield reflections; and this shows at least that the supposed cause is a possible one (Deschanel, *Optics,* § 726). Similarly as to the vastest of all phenomena, the evolution of the stellar system, and of the solar system as part of it, from an immense cloudlike volume of matter: H. Spencer, in his Essay on *The Nebular Hypothesis*, says, amidst a great array of deductive arguments from mechanical principles, that "this *a priori* reasoning harmonises with the results of experiment. Dr. Plateau has shown that when a mass of fluid is, as far as may be, protected from the action of external forces, it will, if made to rotate with adequate velocity, form detached rings; and that these rings will break up into spheroids, which turn on their axes in the same direction with the central mass." The theory of the evolution of species of plants and animals by Natural Selection, again, though, of course, it cannot be verified by direct experiment (since experiment implies artificial arrangement), and the process is too slow for observation, is, nevertheless, to some extent confirmed by the practice of gardeners and breeders of animals: since, by taking advantage of accidental variations of form and colour in the plants or animals under their care, and relying on the inheritability of these variations they obtain extensive modifications of the original stocks, and adapt them to the various purposes for which flowers and cereals, poultry, dogs and cattle are domesticated. This shows, at least, that living forms are plastic, and extensively modifiable in a comparatively short time.

§ 4. Suppose, however, that, in verifying a deductive argument, the effect as computed from the laws of the causes assigned, does not correspond with the facts observed: there must then be an error somewhere. If the fact has been accurately observed, the error must lie either in the process of deduction and computation, or else in the premises. As to the process of deduction, it may be very simple and easily revised, as in the above explanation of the common pump; or it may be very involved and comprise long trains of mathematical calculation. If, however, on re-examining the computations, we find them correct, it remains to look for some mistake in the premises.

(1) We may not have accurately ascertained the laws, or the modes of operation, or the amounts of the forces present. Thus, the rate at which bodies fall was formerly believed to vary in proportion to their relative weights; and any estimate based upon this belief cannot agree with the facts. Again, the corpuscular theory of light, namely, that the physical cause of light is a stream of fine particles projected in straight lines from the luminous object, though it seemed adequate to the explanation of many optical phenomena, could not be made to agree with the facts of interference and double refraction.

(2) The circumstances in which the agents are combined may not have been correctly conceived. When Newton began to inquire whether the attraction of the earth determined the orbit of the moon, he was at first disappointed. "According to Newton's calculations, made at this time," says Whewell, "the moon, by her motion in her orbit, was deflected from the tangent every minute through a space of thirteen feet. But by noticing the space which bodies would fall in one minute at the earth's surface, and supposing this to be diminished in the ratio of the inverse square, it appeared that gravity would, at the moon's orbit, draw a body through more than fifteen feet." In view of this discrepancy he gave up the inquiry for sixteen years, until in 1682, having obtained better data, he successfully renewed it. "He had been mistaken in the magnitude of the earth, and consequently in the distance of the moon, which is determined by measurements of which the earth's radius is the base." It was not, therefore, a mistake as to the law or as to the nature of the forces concerned (namely, the law of the inverse square and the identity of celestial with terrestrial gravity), but as to the circumstances in which the agents (earth and moon) were combined, that prevented his calculations being verified. (*Hist. Ind. Sc.*: VII. ii. 3.)

(3) One or more of the agents affecting the result may have been overlooked and omitted from the estimate. Thus, an attempt to explain the tides by taking account only of the earth and the moon, will not entirely agree with the facts, since the sun also influences the tides. This illustration, however, shows that when the conclusion of a deductive explanation does not entirely agree with the facts, it is not always to be inferred that the reasoning is, properly speaking, wrong; it may be right as far as it goes, and merely inadequate. Hence (*a*) in such cases an opportunity occurs of applying the Method of Residues, by discovering the agent that must be allowed for in order to complete the explanation. And (*b*) the investigation of a phenomenon is often designedly begun upon an imperfect basis for the sake of simplicity; the result being regarded as a first approximation, to be afterwards corrected by including, one by one, the remaining agents or circumstances affecting the phenomenon, until the theory is complete; that is, until its agreement with the facts is satisfactory.

(4) We may have included among the data of our reasonings agents or circumstances that do not exist or do not affect the phenomenon in question. In the early days of science purely fanciful powers were much relied upon: such as the solid spheres that carried the planets and stars; the influence of the planets upon human destiny; the tendency of everything to seek "its own place," so that fire rises to heaven, and solids fall to the earth; the "plastic virtue" of the soil, which was once thought to have produced fossils. When, however, such conceptions hindered the progress of explanation, it was not so much by vitiating the deductive method as by putting men off from exact inquiries. More to our present purpose were the supposed cataclysms, or extraordinary convulsions of the earth, a belief in which long hindered the progress of Geology. Again, in Biology, Psychology, and Sociology many explanations have depended upon the doctrine that any improvement of structure or faculty acquired by an individual may be inherited by his descendants: as that, if an animal learns to climb trees, his offspring have a greater aptitude for that mode of life; that if a man tries to be good, his children find it easier to be virtuous; that if the inhabitants of a district carry on cloth-work, it becomes easier for each successive generation to acquire dexterity in that art. But now

the inheritability of powers acquired by the individual through his own efforts, is disputed; and, if the denial be made good, all such explanations as the above must be revised.

If, then, the premises of a deductive argument be vitiated in any of these four ways, its conclusion will fail to agree with the results of observation and experiment, unless, of course, one kind of error happen to be cancelled by another that is 'equal and opposite.' We now come to a variation of the method of combining Induction with Deduction, so important as to require separate treatment.

§ 5. The Inverse or Historical Method has of late years become remarkably fruitful. When the forces determining a phenomenon are too numerous, or too indefinite, to be combined in a direct deduction, we may begin by collecting an empirical law of the phenomenon (as that 'the democracies of City-States are arbitrary and fickle'), and then endeavour to show by deductions from "the nature of the case," that is, from a consideration of the circumstances and forces known to be operative (of which, in the above instance, the most important is sympathetic contagion), that such a law was to be expected. Deduction is thus called in to verify a previous induction; whereas in the 'Physical Method' a deduction was verified by comparing it with an induction or an experiment; hence the method now to be discussed has been named the Inverse Deductive Method.

But although it is true that, in such inquiries as we are now dealing with, induction generally takes the lead; yet I cannot think that the mere order in which the two logical processes occur is the essential distinction between the two ways of combining them. For, in the first place, in investigations of any complexity both induction and deduction recur again and again in whatever order may be most convenient; and, in the second place, the so-called 'inverse order' is sometimes resorted to in Astronomy and Physics. For example, Kepler's Laws were first collected empirically from observations of the planetary motions, and afterwards deduced by Newton from the Law of Gravitation; this, then, was the Inverse Method; but the result is something very different from any that can be obtained by the Historical Method. The essential difference between the Physical and Historical Methods is that, in the former, whether Direct or Inverse, the deductive process, when complete, amounts to exact demonstration; whereas, in the latter, the deductions may consist of qualitative reasonings, and the results are indefinite. They establish—(1) a merely probable connection between the phenomena according to an empirical law (say, between City-democracy and fickle politics); (2) connect this with other historical or social generalisations, by showing that they all alike flow from the same causes, namely, from the nature of races of men under certain social and geographical conditions; and (3) explain why such empirical laws may fail, according to the differences that prevail among races of men and among the conditions under which they live. Thus, seeing how rapidly excitement is propagated by the chatter, grimacing, and gesticulation of townsmen, it is probable enough that the democracy of a City-state should be fickle (and arbitrary, because irresponsible). A similar phenomenon of panic, sympathetic hope and despair, is exhibited by every stock-exchange, and is not peculiar to political life. And when political opinion is not manufactured solely in the reverberating furnace of a city, fickleness ceases to characterise democracy; and, in fact, is not found in Switzerland, or the United States, nor in France so far as politics, depend upon the peasantry.

This is called the Historical Method, then, because it is especially useful in explaining the movements of history, and in verifying the generalisations of political and social science. We must not, however, suppose that its use is confined to such studies. Only a ridiculous pedantry would allot to each subject its own method and forbid the use of any other; as if it were not our capital object to establish truth by any means. Wherever the forces determining a phenomenon are too numerous or too indefinite to be combined in a deductive demonstration, there the Historical Method is likely to be useful; and this seems often to be the case in Geology and Biology, as well as in the Science of History, or Sociology, and its various subsidiary studies.

Consider upon what causes historical events depend: the customs, character, and opinions of all the people concerned; the organisation of their government, and the character of their religious institutions; the development of industry among them, of the military art, of fine art, literature and science; their relations, commercial, political and social, with other nations; the physical conditions of climate and geographical position amidst which they live. Hardly an event of importance occurs in any nation that is not, directly or indirectly, influenced by every one of these circumstances, and that does not react upon them. Now, from the nature of the Canons of direct Induction, a satisfactory employment of them in such a complex and tangled situation as history presents, is rarely possible; for they all require the actual or virtual isolation of the phenomenon under investigation. They also require the greatest attainable immediacy of connection between cause and effect; whereas the causes of social events may accumulate during hundreds of years. In collecting empirical laws from history, therefore, only very rough inductions can be hoped for, and we may have to be content with simple enumeration. Hence the importance of supporting such laws by deduction from the nature of the case, however faint a probability of the asserted connection is thereby raised; and this even if each law is valued merely for its own sake. Still more, if anything worth the name of Historical Science is to be constructed, must a mere collection of such empiricisms fail to content us; and the only way to give them a scientific character is to show deductively their common dependence upon various combinations of the same causes. Yet even those who profess to employ the Historical Method often omit the deductive half of it; and of course 'practical politicians' boast of their entire contentment with what they call 'the facts.'

Sometimes, however, politicians, venturing upon deductive reasoning, have fallen into the opposite error of omitting to test their results by any comparison with the facts: arguing from certain 'Rights of Man,' or 'Interests of Classes,' or 'Laws of Supply and Demand,' that this or that event will happen, or ought to happen, without troubling themselves to observe whether it does happen or ever has happened. This method of Deduction without any empirical verification, is called by Mill the Geometrical; and, plainly, it can be trustworthy only where there is no actual conflict of forces to be considered. In pure mathematical reasoning about space, time, and number, provided the premises and the reasoning be correct, verification by a comparison with the facts may be needless, because there is no possibility of counteraction. But when we deal with actual causes, no computation of their effects can be relied upon without comparing our conclusions with the facts: not even in Astronomy and Physics, least of all in Politics.

Burke, then, has well said that "without the guide and light of sound, well-understood principles all our reasoning in politics, as in everything else, would be only a confused jumble of particular facts and details without the means of drawing any sort of theoretical or practical conclusion"; but that, on the other hand, the statesman, who does not take account of circumstances, infinite and infinitely combined, "is not erroneous, but stark mad—he is metaphysically mad" (*On the Petition of the Unitarians*). There is, or ought to be, no logical difference between the evidence required by a statesman and that appealed to by a philosopher; and since, as we have seen, the combination of principles with circumstances cannot, in solving problems of social science, be made with the demonstrative precision that belongs to astronomical and physical investigations, there remains the Historical Method as above described.

Examples of the empirical laws from which this method begins abound in histories, newspapers, and political discussions, and are of all shades of truth or half-truth: as that 'History consists in the biographies of great men'; in other words, that the movements of society are due to exceptional personal powers, not to general causes; That at certain epochs great men occur in groups; That every Fine Art passes through periods of development, culmination and decline; That Democracies tend to change into Despotisms; That the possession of power, whether by classes or despots, corrupts the possessor: That 'the governments most distinguished for sustained vigour and abilities have generally been aristocracies'; That 'revolutions always begin in hunger'; That civilisation is inimical to individuality; That the civilisation of the country proceeds from the town; That 'the movement of progressive societies has hitherto been a movement from *Status* to *Contract* (*i.e.*, from a condition in which the individual's rights and duties depend on his caste, or position in his family as slave, child, or patriarch, to a condition in which his rights and duties are largely determined by the voluntary agreements he enters into)'; and this last is treated by H. Spencer as one aspect of the law first stated by Comte, that the progress of societies is from the military to the industrial state.

The deductive process we may illustrate by Spencer's explanation of the co-existence in the military state of those specific characters, the inductive proof of which furnished an illustration of the method of Agreement (ch. xvi. § 1). The type of the military State involves the growth of the warrior class, and the treatment of labourers as existing solely to support the warriors; the complete subordination of all individuals to the will of the despotic soldier-king, their property, liberty and life being at the service of the State; the regimentation of society, not only for military, but also for civil purposes; the suppression of all private associations, *etc*. Now all these characteristics arise from their utility for the purpose of war, a utility amounting to necessity if war is the State's chief purpose. For every purpose is best served when the whole available force co-operates toward it: other things equal, the bigger the army the better; and to increase it, men must be taken from industry, until only just enough remain to feed and equip the soldiers. As this arrangement is not to everybody's taste, there must be despotic control; and this control is most effective through regimentation by grades of command. Private associations, of course, cannot live openly in such a State, because they may have wills of their own and are convenient for conspiracy. Thus the induction of characteristics is verified by a deduction of them from the nature of the case.

§ 6. The greater indefiniteness of the Historical compared with the Physical Method, both in its inductions and in its deductions, makes it even more difficult to work with. It wants much sagacity and more impartiality; for the demon of Party is too much with us. Our first care should be to make the empirical law as nearly true as possible, collecting as many as we can of the facts which the law is supposed to generalise, and examining them according to the canons of Induction, with due allowance for the imperfect applicability of those canons to such complex, unwieldy, and indefinite instances. In the examples of such laws given above, it is clear that in some cases no pains have been taken to examine the facts. What is the inductive evidence that Democracies change into Despotisms; that revolutions always begin in hunger; or that civilisation is inimical to individuality? Even Mill's often quoted saying, "that the governments remarkable in history for sustained vigour and ability have generally been aristocracies," is oddly over-stated. For if you turn to the passage (*Rep. Gov.* chap. vi.), the next sentence tells you that such governments have always been aristocracies of public functionaries; and the next sentence but one restricts, apparently, the list of such remarkable governments to two—Rome and Venice. Whence, then, comes the word "generally" into Mill's law?

As to deducing our empirical law from a consideration of the nature of the case, it is obvious that we ought—(*a*) to take account of all the important conditions; (*b*) to allow weight to them severally in proportion to their importance; and (*c*) not to include in our estimates any condition which we cannot show to be probably present and operative. Thus the Great-Man-Theory of history must surely be admitted to assign a real condition of national success. The great man organises, directs, inspires: is that nothing? On the other hand, to recognise no other condition of national success is the manifest frenzy of a mind in the mythopœic age. We must allow the great man his due weight, and then inquire into the general conditions that (*a*) bring him to birth in one nation rather than another, and (*b*) give him his opportunity.

Mill's explanation of the success of the aristocratic governments of Rome and Venice is, that they were, in fact, bureaucracies; that is to say, their members were trained in the science and art of administration and command. Here, again, we have, no doubt, a real condition; but is it the only one? The popular mind, which little relishes the scaling down of Mill's original law to those two remote cases, is persuaded that an aristocracy is the depository of hereditary virtue, especially with reference to government, and would at once ascribe to this circumstance the greater part of the success of any aristocratic constitution. Now, if the effects of training are inherited, they must, in an hereditary aristocracy, increase the energy of the cause assigned by Mill; but, if not, such heredity is a condition "not present or not operative." Still, if families are ennobled for their extraordinary natural powers of administration or command (as sometimes happens), it is agreed on all hands that innate qualities are inheritable; at least, if care be taken to intermarry with families similarly distinguished, and if by natural or artificial selection all the failures among the offspring be eliminated. The Spartans had some crude notion of both these precautions; and if such measures had been widely adopted, we might deduce from the doctrine of heredity a probability in favour of Mill's original proposition, and thereby verify it in its generality, if it could be collected from the facts.

The Historical Method may be further illustrated by the course adopted in that branch of Social Science which has been found susceptible of the most extensive independent development, namely, Economics. First, by way of contrast, I should say that the abstract, or theoretical treatment of Economics follows the Physical Method; because, as Mill explains, although the phenomena of industry are no doubt influenced, like other social affairs, by all the other circumstances of Society, government, religion, war, art, *etc*.; yet, where industry is most developed, as in England and the United States, certain special conditions affecting it are so much the most important that, for the purpose at least of a first outline of the science, they may conveniently be considered as the only ones. These conditions are: (1) the general disposition of men to obtain wealth with as little trouble as possible, and (2) to spend it so as to obtain the greatest satisfaction of their various desires; (3) the facts that determine population; and (4) the tendency of extractive industry, when pushed beyond a certain limit without any improvement in the industrial arts, to yield "diminishing returns." From these premises it is easy to infer the general laws of prices, of wages and interest (which are the prices of labour and of the use of capital), and of rent; and it remains to verify these laws by comparing them with the facts in each case; and (if they fail to agree with the facts) to amend them, according to the Method of Residues, by taking account of those influential conditions which were omitted from the first draft of the theory.

Whilst, however, this is usually the procedure of those inquirers who have done most to give Economics its scientific character, to insist that no other plan shall be adopted would be sheer pedantry; and Dr. Keynes has shown, in his *Scope and Method of Political Economy*, that Mill has himself sometimes solved economic problems by the Historical Method. With an analysis of his treatment of Peasant Proprietorship (*Political Economy*, B. II., cc. 7 and 8) we may close this section. Mill first shows inductively, by collecting evidence from Switzerland, Germany, Norway, Belgium, and France (countries differing in race, government, climate and situation), that peasant proprietors are superhumanly industrious; intelligent cultivators, and generally intelligent men; prudent, temperate, and independent, and that they exercise self-control in avoiding improvident marriages. This group of empirical generalisations as to the character of peasant proprietors he then deduces from the nature of the case: their industry, he says, is a natural consequence of the fact that, however much they produce, it is all their own; they cultivate intelligently, because for generations they have given their whole mind to it; they are generally intelligent men, because the variety of work involved in small farming, requiring foresight and calculation, necessarily promotes intelligence; they are prudent, because they have something to save, and by saving can improve their station and perhaps buy more land; they are temperate, because intemperance is incompatible with industry and prudence; they are independent, because secure of the necessaries of life, and from having property to fall back upon; and they avoid improvidence in marriage, because the extent and fertility of their fields is always plainly before them, and therefore how many children they can maintain is easily calculated. The worst of them is that they work too hard and deny themselves too much: but, over the greater part of the world, other peasantry work too hard; though they can scarcely be said to deny themselves too much; since all their labour for others brings them no surplus to squander upon self-indulgence.

§ 7. The foregoing account of the Historical Method is based upon Mill's discussions in B. VI. of his *Logic*, especially cc. 6 to 11. Mill ascribes to Comte the first clear statement of the method; and it is highly scientific, and important in generalising the connections of historical events. But perhaps the expression, 'Historical Method,' is more frequently applied to the Comparative Method, as used in investigating the history of institutions or the true sense of legends.

(1) Suppose we are trying to explain the institution of capital punishment as it now exists in England. (1) We must try to trace the history of it back to the earliest times; for *social custom and tradition is one line of causation*. At present the punishment of death is legally incident only to murder and high treason. But early in the last century malefactors were hung for forgery, sheep-stealing, arson and a long list of other offences down to pocket-picking: earlier still the list included witchcraft and heresy. At present hanging is the only mode of putting a malefactor to death; but formerly the ways of putting to death included also burning, boiling, pressing, beheading, and mixed modes. Before the Restoration, however, the offences punishable with death were far fewer than they afterwards became; and until the twelfth century, the penalty of death might be avoided by paying compensation, the wer-geld.

(2) Every change in the history of an institution must be explained by pointing to *the special causes* in operation during the time when the change was in progress. Thus the restriction of the death penalty, in the nineteenth century, to so few offences was due partly to the growth of humane feelings, partly to the belief that the infliction, or threat, of the extreme penalty had failed to enforce the law and had demoralised the administration of Justice. The continual extension of the death penalty throughout the eighteenth century may be attributed to a belief that it was the most effectual means of deterring evil-doers when the means of detecting and apprehending criminals were feeble and ill-organised. The various old brutal ways of execution were adopted sometimes to strike terror, sometimes for vengeance, sometimes from horror of the crime, or even from 'conscientious scruples';—which last were the excuse for preferring the burning of heretics to any sort of bloodshed.

(3) The causes of any change in the history of an institution in any country may not be directly discoverable: they must then be investigated by the Comparative Method. Again, the recorded history of a nation, and of all its institutions, followed backwards, comes at last to an end: then the antecedent history must also be supplied by the Comparative Method; whose special use is to indicate the existence of facts for which there is no direct evidence.

This method rests upon the principle that where the causes are alike the effects will be alike, and that similar effects are traceable to similar causes. Every department of study—Astronomy, Chemistry, Zoology, Sociology—is determined by the fact that the phenomena it investigates have certain common characteristics; and we are apt to infer that any process observed in some of these phenomena, if depending on those common characteristics, will be found in others. For example, the decomposition, or radio-activity, of certain elements prepares one to believe that all elements may exhibit it. Where the properties of an object are known to be closely interdependent, as in the organisation of plants, animals and societies, we are especially justified in inferring from one case to another. The whole animal Kingdom has certain common characters—the metabolic process, dependence upon oxygen, upon vegetable food (ultimately), heredity, etc., and, upon this ground, any process (say, the differentiation of species by Natural Selection) that has been established for some kinds of animal is readily extended to others. If instead of the whole animal Kingdom we take some district of it—Class, Order, Family—our confidence in such inferences increases; because the common characters are more numerous and the conditions of life are more alike; or, in other words, the common causes are more numerous that initiate and control the development of nearly allied animals. For such reasons a few fragmentary remains of an extinct animal enable the palæontologist to reconstruct with some probability an outline of its appearance, organisation, food, habitat and habits.

Applied to History, the Comparative Method rests upon an assumption (which the known facts of (say) 6,000 years amply justify) that human nature, after attaining a recognisable type as *homo sapiens*, is approximately uniform in all countries and in all ages, though more especially where states of culture are similar. Men living in society are actuated by similar motives and reasons in similar ways; they are all dependent upon the supply of food and therefore on the sun and the seasons and the weather and upon means of making fire, and so on. Accordingly, they entertain similar beliefs, and develop similar institutions through similar series of changes. Hence, if in one nation some institution has been altered for reasons that we cannot directly discover, whereas we know the reasons why a similar change was adopted elsewhere, we may conjecture with more or less probability, after making allowance for differences in other circumstances, that the motives or causes in the former case were similar to those in the latter, or in any cases that are better known. Or, again, if in one nation we cannot trace an institution beyond a certain point, but can show that elsewhere a similar institution has had such or such an antecedent history, we may venture to reconstruct with more or less probability the earlier history of that institution in the nation we are studying.

Amongst the English and Saxon tribes that settled in Britain, death was the penalty for murder, and the criminal was delivered to the next-of-kin of his victim for execution; he might, however, compound for his crime by paying a certain compensation. Studying the history of other tribes in various parts of the world, we are able, with much probability, to reconstruct the antecedents of this death-penalty in our own prehistoric ages, and to trace it to the blood-feud; that is, to a tribal condition in which the next-of-kin of a murdered man was socially and religiously bound to avenge him by slaying the murderer or one of his kindred. This duty of revenge is sometimes (and perhaps was at first everywhere) regarded as necessary to appease the ghost of the victim; sometimes as necessary to compensate the surviving members of his family. In the latter case, it is open to them to accept compensation in money or cattle, *etc.* Whether the kin will be ready to accept compensation must depend upon the value they set upon wealth in comparison with revenge; but for the sake of order and tribal strength, it is the interest of the tribe, or its elders, or chieftain, to encourage or even to enforce such acceptance. It is also their interest to take the questions—whether a crime has been committed, by whom, and what compensation is due—out of the hands of the injured party, and to submit them to some sort of court or judicial authority. At first, following ancient custom as much as possible, the act of requital, or the choice of accepting compensation, is left to the next-of-kin; but with the growth of central power these things are entrusted to ministers of the Government. Then revenge has undergone its full transformation into punishment. Very likely the wrong itself will come to be treated as having been done not to the kindred of the murdered man, but to the State or the King, as in fact a "breach of the King's peace." This happened in our own history.

(4) The Comparative Method assumes that human nature is approximately the same in different countries and ages; but, of course, 'approximately' is an important word. Although there is often a striking and significant resemblance between the beliefs and institutions of widely separated peoples, we expect to draw the most instructive parallels between those who are nearly related by descent, or neighbourhood, or culture. To shed light upon our own manners, we turn first to other Teutons, then to Slavonians and Kelts, or other Aryans, and so on; and we prefer evidence from Europe to examples from Africa.

(5) As to national culture, that it exhibits certain 'stages' of development is popularly recognised in the distinction drawn between savages, barbarians and civilised folk. But the idea remains rather vague; and there is not space here to define it. I refer, therefore, to the classifications of stages of culture given by A. Sutherland, (*Origin and Growth of Moral Instinct*, Vol. I, p. 103), and L.T. Hobhouse (*Morals in Evolution*, c. 2). That in any 'state of Society,' its factors—religion, government, science, *etc.*—are mutually dependent, was a leading doctrine with Comte, adopted by Mill. There must be some truth in it; but in some cases we do not understand social influences sufficiently well to trace the connection of factors; and whilst preferring to look for historical parallels between nations of similar culture, we find many cases in which barbarous or savage customs linger in a civilised country.

(6) It was another favourite doctrine with Comte, also adopted by Mill—that the general state of culture is chiefly determined by the prevailing intellectual condition of a people, especially by the accepted ground of explanation—whether the will of supernatural beings, or occult powers, or physical antecedents: the "law of three stages," Fetichism, Metaphysics, Positivism. And this also is, at least, so far true, that it is useless to try to interpret the manners and institutions of any nation until we know its predominant beliefs. Magic and animism are beliefs everywhere held by mankind in early stages of culture, and they influence every action of life. But that is not all: these beliefs retain their hold upon great multitudes of civilised men and affect the thoughts of the most enlightened. Whilst the saying 'that human nature is the same in all ages' seems to make no allowance for the fact that, in some nations, a considerable number of individuals has attained to powers of deliberation, self-control, and exact reasoning, far above the barbarous level, it is yet so far true that, even in civilised countries, masses of people, were it not for the example and instruction of those individuals, would fall back upon magic and animism and the manners that go with those beliefs. The different degrees of enlightenment enjoyed by different classes of the population often enable the less educated to preserve a barbarous custom amidst many civilised characteristics of the national life.

§ 8. Historical reasoning must start from, or be verified by, observations. If we are writing the history of ourselves: if of another time or country, we can observe some of the present conditions of the country, its inhabitants, language, manners, institutions, which are effects of the past and must be traceable to it; we may also be able to observe ancient buildings or their ruins, funerary remains, coins, dating from the very times we are to treat of. Our own observations, of course, are by no means free from error.

But even in treating of our own age and country, most of our information must be derived from the testimony of others, who may have made mistakes of observation and further mistakes in reporting their observations, or may have intentionally falsified them. Testimony is of two kinds: Oral; and Written, inscribed or printed. In investigating the events of a remote age, nearly all our direct evidence must be some sort of testimony.

(1) Oral testimony depends upon the character of the witness; and the best witness is not perfectly trustworthy; for he may not have observed accurately, or he may not have reported correctly; especially if some time elapsed between the event and his account of it; for no man's memory is perfect. Since witnesses vary widely in capacity and integrity, we must ask concerning any one of them—was he a good judge of what he saw, and of what was really important in the event? Had he good opportunities of knowing the circumstances? Had he any interest in the event—personal, or partisan, or patriotic? Such interests would colour his report; and so would the love of telling a dramatic story, if that was a weakness of his. Nay, a love of truth might lead him to modify the report of what he remembered if—as he remembered it—the matter seemed not quite credible. We must also bear in mind that, for want of training, precision in speaking the truth is not understood or appreciated by many honest people even now, still less in unscientific ages.

Oral tradition is formed by passing a report from one to another, generation by generation; and it is generally true that such a tradition loses credit at every step, because every narrator has some weakness. However, the value of tradition depends upon the motives people have to report correctly, and on the form of the communication, and on whether monuments survive in connection with the story. Amongst the things best remembered are religious and magic formulæ, heroic poems, lists of ancestors, popular legends about deeply impressive events, such as migrations, conquests, famines, plagues. We are apt now to underrate the value of tradition, because the use of writing has made tradition less important, and therefore less pains are taken to preserve it. In the middle of last century, it was usual (and then quite justifiable) to depreciate oral tradition as nearly worthless; but the spread of archæological and anthropological research, and the growth of the Comparative Method, have given new significance to legends and traditions which, merely by themselves, could not deserve the slightest confidence.

(2) As to written evidence, contemporary inscriptions—such as are found on rocks and stones and bricks in various parts of the world, and most abundantly in Egypt and Western Asia—are of the highest value, because least liable to fraudulent abuse; but must be considered with reference to the motives of those who set them forth. Manuscripts and books give rise to many difficulties. We have to consider whether they were originally written by some one contemporary with the events recorded: if so they have the same value as immediate oral testimony, provided they have not been tampered with since. But if not contemporary records, they may have been derived from other records that were contemporary, or only from oral tradition. In the latter case they are vitiated by the weakness of oral tradition. In the former case, we have to ask what was the trustworthiness of the original records, and how far do the extant writings fairly represent those records?

Our answers to these questions will partly depend upon what we know or can discover of the authors of the MSS. or books. Who was the author? If a work bears some man's name, did he really write it? The evidence bearing upon this question is usually divided into internal, external and mixed; but perhaps no evidence is purely internal, if we define it as that which is derived entirely from the work itself. Under the name of internal evidence it is usual to put the language, the style, consistency of ideas; but if we had no grounds of judgment but the book itself, we could not possibly say whether the style was the author's: this requires us to know his other works. Nor could we say whether the language was that of his age, unless we knew other literature of the same age; nor even that different passages seem to be written in the manner of different ages, but for our knowledge of change in other literatures. There must in every case be some external reference. Thus we judge that a work is not by the alleged author, nor contemporary with him, if words are used that only became current at a later date, or are used in a sense that they only later acquired, or if later writers are imitated, or if events are mentioned that happened later ('anachronism'). Books are sometimes forged outright, that is, are written by one man and deliberately fathered upon another; but sometimes books come to be ascribed to a well-known name, which were written by some one else without fraudulent intent, dramatically or as a rhetorical exercise.

As to external evidence, if from other sources we have some knowledge of the facts described in a given book, and if it presents no serious discrepancies with those facts, this is some confirmation of a claim to contemporaneity. But the chief source of external evidence is other literature, where we may find the book in question referred to or quoted. Such other literature may be by another author, as when Aristotle refers to a dialogue of Plato's, or Shakespeare quotes Marlowe; or may be other work of the author himself, as when Aristotle in the *Ethics* refers to his own *Physics*, or Chaucer in *The Canterbury Tales* mentions as his own *The Legend of Good Women*, and in *The Legend* gives a list of other works of his. This kind of argument assumes that the authorship of the work we start from is undisputed; which is practically the case with the *Ethics* and *The Canterbury Tales*.

But, now, granting that a work is by a good author, or contemporary with the events recorded, or healthily related to others that were contemporary, it remains to consider whether it has been well preserved and is likely to retain its original sense. It is, therefore, desirable to know the history of a book or MS., and through whose hands it has passed. Have there been opportunities of tampering with it; and have there been motives to do so? In reprinting books, but still more in copying MSS., there are opportunities of omitting or interpolating passages, or of otherwise altering the sense. In fact, slight changes are almost sure to be made even without meaning to make them, especially in copying MSS., through the carelessness or ignorance of transcribers. Hence the oldest MS. is reckoned the best.

If a work contains stories that are physically impossible, it shows a defect of judgment in the author, and decreases our confidence in his other statements; but it does not follow that these others are to be rejected. We must try to compare them with other evidence. Even incredible stories are significant: they show what people were capable of believing, and, therefore, under what conditions they reasoned and acted. One cause of the incredibility of popular stories is the fusion of legend with myth. A legend is a traditionary story about something that really happened: it may have been greatly distorted by stupidity, or exaggeration, or dramatisation, or rationalisation, but may still retain a good deal of the original fact. A myth, however, has not necessarily any basis of fact: it may be a sort of primitive philosophy, an hypothesis freely invented to explain some fact in nature, such as eclipses, or to explain some social custom whose origin is forgotten, such as the sacrificing of a ram.

All historical conclusions, then, depend on a sum of convergent and conflicting probabilities in the nature of circumstantial evidence. The best testimony is only highly probable, and it is always incomplete. To complete the picture of any past age there is no resource but the Comparative Method. We use this method without being aware of it, whenever we make the records of the last generation intelligible to ourselves by our own experience. Without it nothing would be intelligible: an ancient coin or weapon would have no meaning, were we not acquainted with the origins and uses of other coins and weapons. Generally, the further we go back in history, the more the evidence needs interpretation and reconstruction, and the more prominent becomes the appeal to the Comparative Method. Our aim is to construct a history of the world, and of the planet as part of the world, and of mankind as part of the life of the planet, in such a way that every event shall be consistent with, and even required by, the rest according to the principle of Causation.

CHAPTER XVIII
HYPOTHESES

§ 1. An Hypothesis, sometimes employed instead of a known law, as a premise in the deductive investigation of nature, is defined by Mill as "any supposition which we make (either without actual evidence, or on evidence avowedly insufficient) in order to endeavour to deduce from it conclusions in accordance with facts which are known to be real; under the idea that if the conclusions to which the hypothesis leads are known truths, the hypothesis itself either must be, or at least is likely to be, true." The deduction of known truths from an hypothesis is its Verification; and when this has been accomplished in a good many cases, and there are no manifest failures, the hypothesis is often called a Theory; though this term is also used for the whole system of laws of a certain class of phenomena, as when Astronomy is called the 'theory of the heavens.' Between hypothesis and theory in the former sense no distinct line can be drawn; for the complete proof of any speculation may take a long time, and meanwhile the gradually accumulating evidence produces in different minds very different degrees of satisfaction; so that the sanguine begin to talk of 'the theory,' whilst the circumspect continue to call it 'the hypothesis.'

An Hypothesis may be made concerning (1) an Agent, such as the ether; or (2) a Collocation, such as the plan of our solar system—whether geocentric or heliocentric; or (3) a Law of an agent's operation, as that light is transmitted by a wave motion of such lengths or of such rates of vibration.

The received explanation of light involves both an agent, the ether, as an all-pervading elastic fluid, and also the law of its operation, as transmitting light in waves of definite form and length, with definite velocity. The agreement between the calculated results of this complex hypothesis and the observed phenomena of light is the chief part of the verification; which has now been so successfully accomplished that we generally hear of the 'Undulatory Theory.' Sometimes a new agent only is proposed; as the planet Neptune was at first assumed to exist in order to account for perturbations in the movements of Uranus, influencing it according to the already established law of gravitation. Sometimes the agents are known, and only the law of their operation is hypothetical, as was at first the case with the law of gravitation itself. For the agents, namely, Earth, falling bodies on the Earth, Moon, Sun, and planets were manifest; and the hypothesis was that their motions might be due to their attracting one another with a force inversely proportional to the squares of the distances between them. In the Ptolemaic Astronomy, again, there was an hypothesis as to the collocation of the heavenly bodies (namely, that our Earth was the centre of the universe, and that Moon, Sun, planets and stars revolved around her): in the early form of the system there was also an hypothesis concerning agents upon which this arrangement depended (namely, the crystalline spheres in which the heavenly bodies were fixed, though these were afterwards declared to be imaginary); and an hypothesis concerning the law of operation (namely, that circular motion is the most perfect and eternal, and therefore proper to celestial things).

Hypotheses are by no means confined to the physical sciences: we all make them freely in private life. In searching for anything, we guess where it may be before going to look for it: the search for the North Pole was likewise guided by hypotheses how best to get there. In estimating the characters or explaining the conduct of acquaintances or of public men, we frame hypotheses as to their dispositions and principles. 'That we should not impute motives' is a peculiarly absurd maxim, as there is no other way of understanding human life. To impute bad motives, indeed, when good are just as probable, is to be wanting in the scientific spirit, which views every subject in 'a dry light.' Nor can we help 'judging others by ourselves'; for self-knowledge is the only possible starting-point when we set out to interpret the lives of others. But to understand the manifold combinations of which the elements of character are susceptible, and how these are determined by the breeding of race or family under various conditions, and again by the circumstances of each man's life, demands an extraordinary union of sympathetic imagination with scientific habits of thought. Such should be the equipment of the historian, who pursues the same method of hypothesis when he attempts to explain (say) the state of parties upon the Exclusion Bill, or the policy of Louis XI. Problems such as the former of these are the easier; because, amidst the compromises of a party, personal peculiarities obliterate one another, and expose a simpler scheme of human nature with fewer fig-leaves. Much more hazardous hypotheses are necessary in interpreting the customs of savages, and the feelings of all sorts of animals. Literary criticisms, again, abound with hypotheses: *e.g.*, as to the composition of the Homeric poems, the order of the Platonic dialogues, the authorship of the Cædmonic poems, or the Ossianic, or of the letters of Junius. Thus the method of our everyday thoughts is identical with that of our most refined speculations; and in every case we have to find whether the hypothesis accounts for the facts.

§ 2. It follows from the definition of an hypothesis that none is of any use that does not admit of verification (proof or disproof), by comparing the results that may be deduced from it with facts or laws. If so framed as to elude every attempt to test it by facts, it can never be proved by them nor add anything to our understanding of them.

Suppose that a conjurer asserts that his table is controlled by the spirit of your deceased relative, and makes it rap out an account of some adventure that could not easily have been within a stranger's knowledge. So far good. Then, trying again, the table raps out some blunder about your family which the deceased relative could not have committed; but the conjurer explains that 'a lying spirit' sometimes possesses the table. This amendment of the hypothesis makes it equally compatible with success and with failure. To pass from small things to great, not dissimilar was the case of the Ptolemaic Astronomy: by successive modifications, its hypothesis was made to correspond with accumulating observations of the celestial motions so ingeniously that, until the telescope was invented, it may be said to have been unverifiable. Consider, again, the sociological hypothesis, that civil order was at first founded on a Contract which remains binding upon all mankind: this is reconcilable with the most opposite institutions. For we have no record of such an event: and if the institutions of one State (say the British) include ceremonies, such as the coronation oath and oath of allegiance, which may be remnants of an original contract, they may nevertheless be of comparatively recent origin; whereas if the institutions of another State (say the Russian) contain nothing that admits of similar interpretation, yet traces of the contract once existing may long since have been obliterated. Moreover, the actual contents of the contract not having been preserved, every adherent of this hypothesis supplies them at his own discretion, 'according to the dictates of Reason'; and so one derives from it the duty of passive obedience, and another with equal cogency establishes the right of rebellion.

To be verifiable, then, an hypothesis must be definite; if somewhat vague in its first conception (which is reasonably to be expected), it must be made definite in order to be put to the proof. But, except this condition of verifiability, and definiteness for the sake of verifiability, without which a proposition does not deserve the name of an hypothesis, it seems inadvisable to lay down rules for a 'legitimate' hypothesis. The epithet is misleading. It suggests that the Logician makes rules for scientific inquirers; whereas his business is to discover the principles which they, in fact, employ in what are acknowledged to be their most successful investigations. If he did make rules for them, and they treated him seriously, they might be discouraged in the exercise of that liberty of hypothesising which is the condition of all originality; whilst if they paid no attention to him, he must suffer some loss of dignity. Again, to say that a 'legitimate hypothesis' must explain all the facts, at least in the department for which it is invented, is decidedly discouraging. No doubt it may be expected to do this in the long run when (if ever) it is completely established; but this may take a long time: is it meanwhile illegitimate? Or can this adjective be applied to Newton's corpuscular theory of light, even though it has failed to explain all the facts?

§ 3. Given a verifiable hypothesis, however, what constitutes proof or disproof?

(1) *If a new agent be proposed, it is desirable that we should be able directly to observe it, or at least to obtain some evidence of its existence of a different kind from the very facts which it has been invented to explain.* Thus, in the discovery of Neptune, after the existence of such a planet outside the orbit of Uranus had been conjectured (to account for the movements of the latter), the place in the heavens which such a body should occupy at a certain time was calculated, and there by means of the telescope it was actually seen.

Agents, however, are assumed and reasoned upon very successfully which, by their nature, never can be objects of perception: such are the atoms of Chemistry and the ether of Optics. But the severer methodologists regard them with suspicion: Mill was never completely convinced about the ether; the defining of which has been found very difficult. He was willing, however, to make the most of the evidence that has been adduced as indicating a certain property of it distinct from those by which it transmits radiation, namely, mechanical inertia, whereby it has been supposed to retard the career of the heavenly bodies, as shown especially by the history of Encke's comet. This comet returned sooner than it should, as calculated from the usual data; the difference was ascribed to the influence of a resisting medium in reducing the extent of its orbit; and such a medium may be the ether. If this conjecture (now of less credit) should gain acceptance, the ether might be regarded as a *vera causa* (that is, a condition whose existence may be proved independently of the phenomena it was intended to explain), in spite of its being excluded by its nature from the sphere of direct perception. However, science is not a way of perceiving things, but essentially a way of thinking about them. It starts, indeed, from perception and returns to it, and its thinking is controlled by the analogies of perception. Atoms and ether are thought about as if they could be seen or felt, not as noumena; and if still successful in connecting and explaining perceptions, and free from contradiction, they will stand as hypotheses on that ground.

On the other hand, a great many agents, once assumed in order to explain phenomena, have since been explained away. Of course, a *fact* can never be 'explained away': the phrase is properly applicable to the fate of erroneous hypotheses, when, not only are they disproved, but others are established in their places. Of the Aristotelian spheres, which were supposed to support and translate sun, moon and planets, no trace has ever been found: they would have been very much in the way of the comets. Phlogiston, again, an agent much in favour with the earlier Chemists, was found, Whewell tells us, when their theories were tested by exact weighing, to be not merely non-existent but a minus quantity; that is to say, it required the assumption of its absolute lightness "so that it diminished the weight of the compounds into which it entered." These agents, then, the spheres and phlogiston, have been explained away, and instead of them we have the laws of motion and oxygen.

(2) *Whether the hypothetical agent be perceptible or not, it cannot be established as a cause, nor can a supposed law of such an agent be accepted as sufficient to the given inquiry, unless it is adequate to account for the effects which it is called upon to explain, at least so far as it pretends to explain them.* The general truth of this is sufficiently obvious, since to explain the facts is the purpose of an hypothesis; and we have seen that Newton gave up his hypothesis that the moon was a falling body, as long as he was unable to show that the amount of its deflection from a tangent (or fall) in a given time, was exactly what it should be, if the Moon was controlled by the same force as falling bodies on the Earth.

It is important to observe the limitations to this canon. In the first place, it says that, unless adequate to explain the facts in question, an hypothesis cannot be '*established*'; but, for all that, such an hypothesis may be a very promising one, not to be hastily rejected, since it may take a very long time fully to verify an hypothesis. Some facts may not be obtainable that are necessary to show the connection of others: as, for example, the hypothesis that all species of animals have arisen from earlier ones by some process of gradual change, can be only imperfectly verified by collecting the fossil remains of extinct species, because immense depths and expanses of fossiliferous strata have been destroyed. Or, again, the general state of culture may be such as to prevent men from tracing the consequences of an hypothesis; for which reason, apparently, the doctrine that the Sun is the centre of our planetary system remained a discredited hypothesis for 2000 years. This should instruct us not to regard an hypothesis as necessarily erroneous or illegitimate merely because we cannot yet see how it works out: but neither can we in such a case regard it as established, unless we take somebody's word for it.

Secondly, the canon says that an hypothesis is not established, unless it accounts for the phenomena *so far as it professes to*. But it implies a complete misunderstanding to assail a doctrine for not explaining what lies beyond its scope. Thus, it is no objection to a theory of the origin of species, that it does not explain the origin of life: it does not profess to. For the same reason, it is no objection to the theory of Natural Selection, that it does not account for the variations which selection presupposes. But such objections might be perfectly fair against a general doctrine of Evolution.

An interesting case in Wallace's *Darwinism* (chap. x.) will illustrate the importance of attending to the exact conditions of an hypothesis. He says that in those groups of "birds that need protection from enemies," "when the male is brightly coloured and the female sits exposed on the nest, she is always less brilliant and generally of quite sober and protective hues"; and his hypothesis is, that these sober hues have been acquired or preserved by Natural Selection, because it is important to the family that the sitting bird should be inconspicuous. Now to this it might be objected that in some birds both sexes are brilliant or conspicuous; but the answer is that the female of such species *does not sit exposed on the nest*; for the nests are either domed over, or made in a hole; so that the sitting bird does not need protective colouring. If it be objected, again, that some sober-coloured birds build domed nests, it may be replied that the proposition 'All conspicuously coloured birds are concealed in the nest,' is not to be converted simply into 'All birds that sit concealed in the nest are conspicuously coloured.' In the cases alleged the domed nests are a protection against the weather, and the sober colouring is a general protection to the bird, which inhabits an open country. It may be urged, however, that jays, crows, and magpies are conspicuous birds, and yet build open nests: but these are aggressive birds, *not needing protection from enemies*. Finally, there are cases, it must be confessed, in which the female is more brilliant than the male, and which yet have open nests. Yes: but *then the male sits upon the eggs*, and the female is stronger and more pugnacious!

Thus every objection is shown to imply some inattention to the conditions of the hypothesis; and in each case it may be said, *exceptio probat regulam*—the exception *tests* the rule. (Of course, the usual translation "proves the rule," in the restricted modern sense of "prove," is absurd.) That is to say, it appears on examination: (1) that the alleged exception is not really one, and (2) that it stands in such relation to the rule as to confirm it. For to all the above objections it is replied that, granting the phenomenon in question (special protective colouring for the female) to be absent, the alleged cause (need of protection) is also absent; so that the proof is, by means of the objections, extended, from being one by the method of Agreement, into one by the Double Method.

Thirdly, an hypothesis originally intended to account for the whole of a phenomenon and failing to do so, though it cannot be established in that sense, may nevertheless contain an essential part of the explanation. The Neptunian Hypothesis in Geology, was an attempt to explain the formation of the Earth's outer crust, as having been deposited from an universal ocean of mud. In the progress of the science other causes, seismic, fluvial and atmospheric, have been found necessary in order to complete the theory of the history of the Earth's crust;

but it remains true that the stratified rocks, and some that have lost their stratified character, were originally deposited under water. Inadequacy, therefore, is not a reason for entirely rejecting an hypothesis or treating it as illegitimate.

(3) Granting that the hypothetical cause is real and adequate, the investigation is not complete. Agreement with the facts is a very persuasive circumstance, the more so the more extensive the agreement, especially if no exceptions are known. Still, if this is all that can be said in favour of an hypothesis, it amounts to proof at most by the method of Agreement; it does not exclude the possibility of vicarious causes; and if the hypothesis proposes a new agent that cannot be directly observed, an equally plausible hypothesis about another imagined agent may perhaps be invented.

According to Whewell, it is a strong mark of the truth of an hypothesis when it agrees with distinct inductions concerning different classes of facts, and he calls this the 'Consilience of Inductions,' because they jump together in the unity of the hypothesis. It is particularly convincing when this consilience takes place easily and naturally without necessitating the mending and tinkering of the hypothesis; and he cites the Theory of Gravitation and the Undulatory Theory of Light as the most conspicuous examples of such ever-victorious hypotheses. Thus, gravitation explains the fall of bodies on the Earth, and the orbits of the planets and their satellites; it applies to the tides, the comets, the double stars, and gives consistency to the Nebular Hypothesis, whence flow important geological inferences; and all this without any need of amendment. Nevertheless, Mill, with his rigorous sense of duty, points out, that an induction is merely a proposition concerning many facts, and that a consilience of inductions is merely a multiplication of the facts explained; and that, therefore, if the proof is merely Agreement in each case, there can be no more in the totality; the possibility of vicarious causes is not precluded; and the hypothesis may, after all, describe an accidental circumstance.

Whewell also laid great stress upon prediction as a mark of a true hypothesis. Thus, Astronomers predict eclipses, occultations, transits, long beforehand with the greatest precision; and the prediction of the place of Neptune by sheer force of deduction is one of the most astonishing things in the history of science. Yet Mill persisted in showing that a predicted fact is only another fact, and that it is really not very extraordinary that an hypothesis, that happens to agree with many known facts, should also agree with some still undiscovered. Certainly, there seems to be some illusion in the common belief in the probative force of prediction. Prediction surprises us, puts us off our guard, and renders persuasion easy; in this it resembles the force of an epigram in rhetoric. But cases can be produced in which erroneous hypotheses have led to prediction; and Whewell himself produces them. Thus, he says that the Ptolemaic theory was confirmed by its predicting eclipses and other celestial phenomena, and by leading to the construction of Tables in which the places of the heavenly bodies were given at every moment of time. Similarly, both Newton's theory of light and the chemical doctrine of phlogiston led to predictions which came true.

What sound method demands in the proof of an hypothesis, then, is *not merely that it be shown to agree with the facts, but that every other hypothesis be excluded*. This, to be sure, may be beyond our power; there may in some cases be no such negative proof except the exhaustion of human ingenuity in the course of time. The present theory of colour has in its favour the failure of Newton's corpuscular hypothesis and of Goethe's anti-mathematical hypothesis; but the field of conjecture remains open. On the other hand, Newton's proof that the solar system is controlled by a central force, was supported by the demonstration that a force having any other direction could not have results agreeing with Kepler's second law of the planetary motions, namely, that, as a planet moves in its orbit, the areas described by a line drawn from the sun to the planet are proportional to the times occupied in the planet's motion. When a planet is nearest to the sun, the area described by such a line is least for any given distance traversed by the planet; and then the planet moves fastest: when the planet is furthest from the sun, the area described by such a line is greatest for an equal distance traversed; and then the planet moves slowest. This law may be deduced from the hypothesis of a central force, but not from any other; the proof, therefore, as Mill says, satisfies the method of Difference.

Apparently, to such completeness of demonstration certain conditions are necessary: the possibilities must lie between alternatives, such as A or not-A, or amongst some definite list of cases that may be exhausted, such as equal, greater or less. He whose hypothesis cannot be brought to such a definite issue, must try to refute whatever other hypotheses are offered, and naturally he will attack first the strongest rivals. With this object in view he looks about for a "crucial instance," that is, an observation or experiment that stands like a cross (sign-post) at the parting of the ways to guide us into the right way, or, in plain words, an instance that can be explained by one hypothesis but not by another. Thus the phases of Venus, similar to those of the Moon, but concurring with great changes of apparent size, presented, when discovered by Galileo, a crucial instance in favour of the Copernican hypothesis, as against the Ptolemaic, so far at least as to prove that Venus revolved around the Sun inside the orbit of the Earth. Foucault's experiment determining the velocity of Light (cited in the last chapter) was at first intended as an *experimentum crucis* to decide between the corpuscular and undulatory theories; and answered this purpose, by showing that the velocity of a beam passed through water was less than it should be by the former, but in agreement with the latter doctrine (Deschanel: § 813).

Perhaps experiments of this decisive character are commonest in Chemistry: chemical tests, says Herschel, "are almost universally crucial experiments." The following is abridged from Playfair (*Encycl. Met., Diss.* III.): The Chemists of the eighteenth century observed that metals were rendered heavier by calcination; and there were two ways of accounting for this: either something had been added in the process, though what, they could not imagine; or, something had been driven off that was in its nature light, namely, phlogiston. To decide between these hypotheses, Lavoisier hermetically sealed some tin in a glass retort, and weighed the whole. He then heated it; and, when the tin was calcined, weighed the whole again, and found it the same as before. No substance, therefore, either light or heavy, had escaped. Further, when the retort was cooled and opened, the air rushed in, showing that some of the air formerly within had disappeared or lost its elasticity. On weighing the whole again, its weight was now found to have increased by ten grains; so that ten grains of air had entered when it was opened. The calcined tin was then weighed separately, and proved to be exactly ten grains heavier than when it was placed in the retort; showing that the ten grains of air that had disappeared had combined with the metal during calcination. This experiment, then, decided against phlogiston, and led to an analysis of common air confirming Priestley's discovery of oxygen.

(4) *An hypothesis must agree with the rest of the laws of Nature; and, if not itself of the highest generality, must be derivable from primary laws* (chap. xix. § 1). Gravitation and the diffusion of heat, light and sound from a centre, all follow the 'law of the inverse square,' and agree with the relation of the radius of a sphere to its surface. Any one who should think that he had discovered a new central force would naturally begin to investigate it on the hypothesis that it conformed to the same law as gravitation or light. A Chemist again, who

should believe himself to have discovered a new element, would expect it to fill one of the vacant places in the Periodic Table. Conformity, in such cases, is strong confirmation, and disagreement is an occasion of misgivings.

A narrower hypothesis, as 'that the toad's ugliness is protective', would be supported by the general theory of protective colouring and figure, and by the still more general theory of Natural Selection, if facts could be adduced to show that the toad's appearance does really deter its enemies. Such an hypothesis resembles an Empirical Law in its need of derivation (chap. xix. §§ 1, 2). If underivable from, or irreconcilable with, known laws, it is a mere conjecture or prejudice. The absolute leviation of phlogiston, in contrast with the gravitation of all other forms of matter, discredited that supposed agent. That Macpherson should have found the Ossianic poems extant in the Gaelic memory, was contrary to the nature of oral tradition; except where tradition is organised, as it was for ages among the Brahmins. The suggestion that xanthochroid Aryans were "bleached" by exposure during the glacial period, does not agree with Wallace's doctrine concerning the coloration of Arctic animals. That our forefathers being predatory, like bears, white variations amongst them were then selected by the advantage of concealment, is a more plausible hypothesis.

Although, then, the consilience of Inductions or Hypotheses is not a sufficient proof of their truth, it is still a condition of it; nonconsilience is a suspicious circumstance, and resilience (so to speak), or mutual repugnance, is fatal to one or all.

§ 4. We have now seen that a scientific hypothesis, to deserve the name, must be verifiable and therefore definite; and that to establish itself as a true theory, it must present some symptom of reality, and be adequate and exclusive and in harmony with the system of experience. Thus guarded, hypotheses seem harmless enough; but some people have a strong prejudice against them, as against a tribe of savages without government, or laws, or any decent regard for vested interests. It is well known, too, that Bacon and Newton disparaged them. But Bacon, in his examples of an investigation according to his own method, is obliged, after a preliminary classification of facts, to resort to an hypothesis, calling it *permissio intellectus, interpretatio inchoata* or *vindemiatio prima*. And Newton when he said *hypotheses non fingo*, meant that he did not deal in fictions, or lay stress upon supposed forces (such as 'attraction'), that add nothing to the law of the facts. Hypotheses are essential aids to discovery: speaking generally, deliberate investigation depends wholly upon the use of them.

It is true that we may sometimes observe a train of events that chances to pass before us, when either we are idle or engaged with some other inquiry, and so obtain a new glimpse of the course of nature; or we may try experiments haphazard, and watch the results. But, even in these cases, before our new notions can be considered knowledge, they must be definitely framed in hypotheses and reobserved or experimented upon, with whatever calculations or precautions may be necessary to ensure accuracy or isolation. As a rule, when inquiring deliberately into the cause of an event, whether in nature or in history, we first reflect upon the circumstances of the case and compare it with similar ones previously investigated, and so are guided by a preconception more or less definite of 'what to look for,' what the cause is likely to be, that is, by an hypothesis. Then, if our preconception is justified, or something which we observe leads to a new hypothesis, either we look for other instances to satisfy the canons of Agreement; or (if the matter admits of experiment) we endeavour, under known conditions according to the canon of Difference, to reproduce the event by means of that which our hypothesis assigns as the cause; or we draw remote inferences from our hypothesis, and try to test these by the Inductive Canons.

If we argue from an hypothesis and express ourselves formally, it will usually appear as the major premise; but this is not always the case. In extending ascertained laws to fresh cases, the minor premise may be an hypothesis, as in testing the chemical constitution of any doubtful substance, such as a piece of ore. Some solution or preparation, A, is generally made which (it is known) will, on the introduction of a certain agent, B, give a reaction, C, if the preparation contains a given substance, X. The major premise is the law of reaction—

Whenever A is X, if treated with B it is C.

The minor premise is an hypothesis that the preparation contains X. An experiment then treats A with B. If C result, a probability is raised in favour of the hypothesis that A is X; or a certainty, if we know that C results on that condition only.

So important are hypotheses to science, that Whewell insists that they have often been extremely valuable even though erroneous. Of the Ptolemaic system he says, "We can hardly imagine that Astronomy could, in its outset, have made so great a progress under any other form." It served to connect men's thoughts on the subject and to sustain their interest in working it out; by successive corrections "to save appearances," it attained at last to a descriptive sort of truth, which was of great practical utility; it also occasioned the invention of technical terms, and, in general digested the whole body of observations and prepared them for assimilation by a better hypothesis in the fulness of time. Whewell even defends the maxim that "Nature abhors a vacuum," as having formerly served to connect many facts that differ widely in their first aspect. "And in reality is it not true," he asks, "that nature *does* abhor a vacuum, and does all she can to avoid it?" Let no forlorn cause despair of a champion! Yet no one has accused Whewell of Quixotry; and the sense of his position is that the human mind is a rather feeble affair, that can hardly begin to think except with blunders.

The progress of science may be plausibly attributed to a process of Natural Selection; hypotheses are produced in abundance and variety, and those unfit to bear verification are destroyed, until only the fittest survive. Wallace, a practical naturalist, if there ever was one, as well as an eminent theorist, takes the same view as Whewell of such inadequate conjectures. Of 'Lemuria,' an hypothetical continent in the Indian Ocean, once supposed to be traceable in the islands of Madagascar, Seychelles, and Mauritius, its surviving fragments, and named from the Lemurs, its characteristic denizens, he says (*Island Life*, chap. xix.) that it was "essentially a provisional hypothesis, very useful in calling attention to a remarkable series of problems in geographical distribution , but not affording the true solution of those problems." We see, then, that 'provisional hypotheses,' or working hypotheses,' though erroneous, may be very useful or (as Whewell says) necessary.

Hence, to be prolific of hypotheses is the first attribute of scientific genius; the first, because without it no progress whatever can be made. And some men seem to have a marked felicity, a sort of instinctive judgment even in their guesses, as if their heads were made according to Nature. But others among the greatest, like Kepler, guess often and are often wrong before they hit upon the truth, and themselves, like Nature, destroy many vain shoots and seedlings of science for one that they find fit to live. If this is how the mind works in scientific inquiry (as it certainly is, with most men, in poetry, in fine art, and in the scheming of business), it is useless to complain. We should rather recognise a place for fools' hypotheses, as Darwin did for "fools' experiments." But to complete the scientific character, there must be great patience, accuracy, and impartiality in examining and testing these conjectures, as well as great ingenuity in devising experiments to that end. The want of these qualities leads to crude work and public failure and brings hypotheses into derision. Not partially and hastily to believe in one's own guesses, nor petulantly or timidly to reject them, but to consider the matter, to suspend judgment, is the moral lesson of science: difficult, distasteful, and rarely mastered.

§ 5. The word 'hypothesis' is often used also for the scientific device of treating an Abstraction as, for the purposes of argument, equivalent to the concrete facts. Thus, in Geometry, a line is treated as having no breadth; in Mechanics, a bar may be supposed absolutely rigid, or a machine to work without friction; in Economics, man is sometimes regarded as actuated solely by love of gain and dislike of exertion. The results reached by such reasoning may be made applicable to the concrete facts, if allowance be made for the omitted circumstances or properties, in the several cases of lines, bars, and men; but otherwise all conclusions from abstract terms are limited by their definitions. Abstract reasoning, then (that is, reasoning limited by definitions), is often said to imply 'the hypothesis' that things exist as their names are defined, having no properties but those enumerated in their definitions. This seems, however, a needless and confusing extension of the term; for an hypothesis proposes an agent, collocation, or law hitherto unknown; whereas abstract reasoning proposes to exclude from consideration a good deal that is well known. There seems no reason why the latter device should not be plainly called an Abstraction.

Such abstractions are necessary to science; for no object is comprehensible by us in all its properties at once. But if we forget the limitations of our abstract data, we are liable to make strange blunders by mistaking the character of the results: treating the results as simply true of actual things, instead of as true of actual things only so far as they are represented by the abstractions. In addressing abstract reasoning, therefore, to those who are unfamiliar with scientific methods, pains should be taken to make it clear what the abstractions are, what are the consequent limitations upon the argument and its conclusions, and what corrections and allowances are necessary in order to turn the conclusions into an adequate account of the concrete facts. The greater the number, variety, and subtlety of the properties possessed by any object (such as human nature), the greater are the qualifications required in the conclusions of abstract reasoning, before they can hold true of such an object in practical affairs.

Closely allied to this method of Abstraction is the Mathematical Method of Limits. In his *History of Scientific Ideas* (B. II. c. 12), Whewell says: "The *Idea of a Limit* supplies a new mode of establishing mathematical truths. Thus with regard to the length of any portion of a curve, a problem which we have just mentioned; a curve is not made up of straight lines, and therefore we cannot by means of any of the doctrines of elementary geometry measure the length of any curve. But we may make up a figure nearly resembling any curve by putting together many short straight lines, just as a polygonal building of very many sides may nearly resemble a circular room. And in order to approach nearer and nearer to a curve, we may make the sides more and more small, more and more numerous. We may then possibly find some mode of measurement, some relation of these small lines to other lines, which is not disturbed by the multiplication of the sides, however far it be carried. And thus we may do what is equivalent to measuring the curve itself; for by multiplying the sides we may approach more and more closely to the curve till no appreciable difference remains. The curve line is the *Limit* of the polygon; and in this process we proceed on the *Axiom* that 'What is true up to the Limit is true at the Limit.'"

What Whewell calls the Axiom here, others might call an Hypothesis; but perhaps it is properly a Postulate. And it is just the obverse of the Postulate implied in the Method of Abstractions, namely, that 'What is true of the Abstraction is true of concrete cases the more nearly they approach the Abstraction.' What is true of the 'Economic Man' is truer of a broker than of a farmer, of a farmer than of a labourer, of a labourer than of the artist of romance. Hence the Abstraction may be called a Limit or limiting case, in the sense that it stands to concrete individuals, as a curve does to the figures made up "by putting together many short straight lines." Correspondingly, the Proper Name may be called the Limit of the class-name; since its attributes are infinite, whereas any name whose attributes are less than infinite stands for a possible class. In short, for logical purposes, a Limit may be defined as any extreme case to which actual examples may approach without ever reaching it. And in this sense 'Method of Limits' might be used as a term including the Method of Abstractions; though it would be better to speak of them generically as 'Methods of Approximation.'

We may also notice the Assumptions (as they may be called) that are sometimes employed to facilitate an investigation, because some definite ground must be taken and nothing better can be thought of: as in estimating national wealth, that furniture is half the value of the houses.

It is easy to conceive of an objector urging that such devices as the above are merely ways of avoiding the actual problems, and that they display more cunning than skill. But science, like good sense, puts up with the best that can be had; and, like prudence, does not reject the half-loaf. The position, that a conceivable case that can be dealt with may, under certain conditions, be substituted for one that is unworkable, is a touchstone of intelligence. To stand out for ideals that are known to be impossible, is only an excuse for doing nothing at all.

In another sense, again, the whole of science is sometimes said to be hypothetical, because it takes for granted the Uniformity of Nature; for this, in its various aspects, can only be directly ascertained by us as far as our experience extends; whereas the whole value of the principle of Uniformity consists in its furnishing a formula for the extension of our other beliefs beyond our actual experience. Transcendentalists, indeed, call it a form of Reason, just because it is presupposed in all knowledge; and they and the Empiricists agree that to adduce material evidence for it, in its full extent, is impossible. If, then, material evidence is demanded by any one, he cannot regard the conclusions of Mathematics and Physical Science as depending on what is itself unproved; he must, with Mill, regard these conclusions as drawn "not from but according to" the axioms of Equality and Causation. That is to say, if the axioms are true, the conclusions are; the material evidence for both the axioms and the conclusions being the same, namely, uncontradicted experience. Now when we say, 'If Nature is uniform, science is true,' the hypothetical character of science appears in the form of the statement. Nevertheless, it seems undesirable to call our confidence in Nature's uniformity an 'hypothesis': it is incongruous to use the same term for our tentative conjectures and for our most indispensable beliefs. 'The Universal Postulate' is a better term for the principle which, in some form or other, every generalisation takes for granted.

We are now sometimes told that, instead of the determinism and continuity of phenomena hitherto assumed by science, we should recognise indeterminism and discontinuity. But it will be time enough to fall in with this doctrine when its advocates produce a new Logic of Induction, and explain the use of the method of Difference and of control experiments according to the new postulates.

CHAPTER XIX
LAWS CLASSIFIED; EXPLANATION; CO-EXISTENCE; ANALOGY

§ 1. Laws are classified, according to their degrees of generality, as higher and lower, though the grades may not be decisively distinguishable.

First, there are Axioms or Principles, that is real, universal, self-evident propositions. They are—(1) real propositions; not, like 'The whole is greater than any of its parts,' merely definitions, or implied in definitions. (2) They are regarded as universally true of phenomena, as far as the form of their expression extends; that is, for example, Axioms concerning quantity are true of everything that is considered in its quantitative aspect, though not (of course) in its qualitative aspect. (3) They are self-evident; that is, each rests upon its own evidence (whatever that may be); they cannot be derived from one another, nor from any more general law. Some, indeed, are more general than others: the Logical Principle of Contradiction, 'if A is B, it is not-B', is true of qualities as well as of quantities; whereas the Axioms of Mathematics apply only to quantities. The Mathematical Axioms, again, apply to time, space, mental phenomena, and matter and energy; whereas the Law of Causation is only true of concrete events in the redistribution of matter and energy: such, at least, is the strict limit of Causation, if we identify it with the Conservation of Energy; although our imperfect knowledge of life and mind often drives us to speak of feelings, ideas, volitions, as causes. Still, the Law of Causation cannot be derived from the Mathematical Axioms, nor these from the Logical. The kind of evidence upon which Axioms rest, or whether any evidence can be given for them, is (as before observed) a question for Metaphysics, not for Logic. Axioms are the upward limit of Logic, which, like all the special sciences, necessarily takes them for granted, as the starting point of all deduction and the goal of all generalisation.

Next to Axioms, come Primary Laws of Nature: these are of less generality than the Axioms, and are subject to the conditions of methodical proof; being universally true only of certain forces or properties of matter, or of nature under certain conditions; so that proof of them by logical or mathematical reasoning is expected, because they depend upon the Axioms for their formal evidence. Such are the law of gravitation, in Astronomy; the law of definite proportions, in Chemistry; the law of heredity, in Biology; and in Psychology, the law of relativity.

Then, there are Secondary Laws, of still less generality, resulting from a combination of conditions or forces in given circumstances, and therefore conceivably derivable from the laws of those conditions or forces, if we can discover them and compute their united effects. Accordingly, Secondary Laws are either—(1) Derivative, having been analysed into, and deduced from, Primary Laws; or (2) Empirical, those that have not yet been deduced (though from their comparatively special and complex character, it seems probable they may be, given sufficient time and ingenuity), and that meanwhile rest upon some unsatisfactory sort of induction by Agreement or Simple Enumeration.

Whether laws proved only by the canon of Difference are to be considered Empirical, is perhaps a question: their proof derives them from the principle of Causation; but, being of narrow scope, some more special account of them seems requisite in relation to the Primary Laws before we can call them Derivative in the technical sense.

Many Secondary Laws, again, are partially or imperfectly Derivative; we can give general reasons for them, without being able to determine theoretically the precise relations of the phenomena they describe. Meteorologists can explain the general conditions of all sorts of weather, but have made little progress toward predicting the actual course of it (at least, for our island): Geologists know the general causes of mountain ranges, but not why they rise just where we find them: Economists explain the general course of a commercial crisis, but not why the great crises recurred at intervals of about ten years.

Derivative Laws make up the body of the exact sciences, having been assimilated and organised; whilst Empirical Laws are the undigested materials of science. The theorems of Euclid are good examples of derivative laws in Mathematics; in Astronomy, Kepler's laws and the laws of the tides; in Physics, the laws of shadows, of perspective, of harmony; in Biology, the law of protective coloration; in Economics, the laws of prices, wages, interest, and rent.

Empirical Laws are such as Bode's law of the planetary distances; the laws of the expansion of different bodies by heat, and formulæ expressing the electrical conductivity of each substance as a function of the temperature. Strictly speaking, I suppose, all the laws of chemical combination are empirical: the law of definite proportions is verifiable in all cases that have been examined, except for variations that may be ascribed to errors of experiment. Much the same is true in Biology; most of the secondary laws are empirical, except so far as structures or functions may be regarded as specialised cases in Physics or Chemistry and deducible from these sciences. The theory of Natural Selection, however, has been the means of rendering many laws, that were once wholly empirical, at least partially derivative; namely, the laws of the geographical distribution of plants and animals, and of their adaptation in organisation, form and colour, habits and instincts, to their various conditions of life. The laws that remain empirical in Biology are of all degrees of generality from that of the tendency to variation in size and in every other character shown by every species (though as to the reason of this there are promising hypotheses), down to such curious cases as that the colour of roses and carnations never varies into blue, that scarlet flowers are never sweet-scented, that bullfinches fed on hemp-seed turn black, that the young of white, yellow and dun pigeons are born almost naked (whilst others have plenty of down); and so on. The derivation of empirical laws is the greater part of the explanation of Nature (§§ 5, 6).

A 'Fact,' in the common use of the word, is a particular observation: it is the material of science in its rawest state. As perceived by a mind, it is, of course, never absolutely particular: for we cannot perceive anything without classing it, more or less definitely, with things already known to us; nor describe it without using connotative terms which imply a classification of the things denoted. Still, we may consider an observation as particular, in comparison with a law that includes it with numerous others in one general proposition. To turn an observation into an experiment, or (where experiment is impracticable) to repeat it with all possible precautions and exactness, and to describe it as to the duration, quantity, quality and order of occurrence of its phenomena, is the first stage of scientific manufacture. Then comes the formulation of an empirical law; and lastly, if possible, deduction or derivation, either from higher laws previously ascertained, or from an hypothesis. However, as a word is used in various senses, we often speak of laws as 'facts': we say the law of gravitation is a fact, meaning that it is real, or verifiable by observations or experiments.

§ 2. Secondary Laws may also be classified according to their constancy into—(1) the Invariable (as far as experience reaches), and (2) Approximate Generalisations in the form—Most X's are Y. Of the invariable we have given examples above. The following are approximate generalisations: Most comets go round the Sun from East to West; Most metals are solid at ordinary temperatures; Most marsupials are Australasian; Most arctic animals are white in winter; Most cases of plague are fatal; Most men think first of their own interests. Some of these laws are empirical, as that 'Most metals are solid at ordinary temperatures': at present no reason can be given for this; nor do we know why most cases of plague are fatal. Others, however, are at least partially derivative, as that 'Most arctic animals are white'; for this seems to be due to the advantage of concealment in the snow; whether, as with the bear, the better to surprise its prey, or, with the hare, to escape the notice of its enemies.

But the scientific treatment of such a proposition requires that we should also explain the exceptions: if 'Most are,' this implies that 'Some are not'; why not, then? Now, if we can give reasons for all the exceptions, the approximate generalisation may be converted into an universal one, thus: 'All arctic animals are white, unless (like the raven) they need no concealment either to prey or to escape; or unless mutual recognition is more important to them than concealment (as with the musk-sheep)'. The same end of universal statement may be gained by including the conditions on which the phenomenon depends, thus: 'All arctic animals to whom concealment is of the utmost utility are white.'

When statistics are obtainable, it is proper to convert an approximate generalisation into a proportional statement of the fact, thus: instead of 'Most attacks of plague are fatal', we might find that in a certain country 70 per cent. were so. Then, if we found that in another country the percentage of deaths was 60, in another 40, we might discover, in the different conditions of these countries, a clue to the high rate of mortality from this disease. Even if the proportion of cases in which two facts are connected does not amount to 'Most,' yet, if any definite percentage is obtainable, the proposition has a higher scientific value than a vague 'Some': as if we know that 2 per cent. of the deaths in England are due to suicide, this may be compared with the rates of suicide in other countries; from which perhaps inferences may be drawn as to the causes of suicide.

In one department of life, namely, Politics, there is a special advantage in true approximate generalisations amounting to 'Most cases.' The citizens of any State are so various in character, enlightenment, and conditions of life, that we can expect to find few propositions universally true of them: so that propositions true of the majority must be trusted as the bases of legislation. If most men are deterred from crime by fear of punishment; if most men will idle if they can obtain support without industry; if most jurymen will refuse to convict of a crime for which the prescribed penalties seem to them too severe; these are most useful truths, though there should be numerous exceptions to them all.

§ 3. Secondary Laws can only be trusted in 'Adjacent Cases'; that is, where the circumstances are similar to those in which the laws are known to be true.

A Derivative Law will be true wherever the forces concerned exist in the combinations upon which the law depends, if there are no counteracting conditions. That water can be pumped to about 33 feet at the sea-level, is a derivative law on this planet: is it true in Mars? That depends on whether there are in Mars bodies of a liquid similar to our water; whether there is an atmosphere there, and how great its pressure is; which will vary with its height and density. If there is no atmosphere there can be no pumping; or if there is an atmosphere of less pressure than ours, water such as ours can only be pumped to a less height than 33 feet. Again, we know that there are arctic regions in Mars; if there are also arctic animals, are they white? That may depend upon whether there are any beasts of prey. If not, concealment seems to be of no use.

An Empirical Law, being one whose conditions we do not know, the extent of its prevalence is still less ascertainable. Where it has not been actually observed to be true, we cannot trust it unless the circumstances, on the whole, resemble so closely those amongst which it has been observed, that the unknown causes, whatever they may be, are likely to prevail there. And, even then, we cannot have much confidence in it; for there may be unknown circumstances which entirely frustrate the effect. The first naturalist who travelled (say) from Singapore eastward by Sumatra and Java, or Borneo, and found the mammalia there similar to those of Asia, may naturally have expected the same thing in Celebes and Papua; but, if so, he was entirely disappointed; for in Papua the mammalia are marsupials like those of Australia. Thus his empirical law, 'The mammalia of the Eastern Archipelago are Asiatic,' would have failed for no apparent reason. According to Mr. Wallace, there is a reason for it, though such as could only be discovered by extensive researches; namely, that the sea is deep between Borneo and Celebes, so that they must have been separated for many ages; whereas it is shallow from Borneo westward to Asia, and also southward from Celebes to Australia; so that these regions, respectively, may have been recently united: and the true law is that similar mammalia belong to those tracts which at comparatively recent dates have formed parts of the same continents (unless they are the remains of a former much wider distribution).

A considerable lapse of time may make an empirical law no longer trustworthy; for the forces from whose combination it resulted may have ceased to operate, or to operate in the same combination; and since we do not know what those forces were, even the knowledge that great changes have taken place in the meantime cannot enable us, after an interval, to judge whether or not the law still holds true. New stars shine in the sky and go out; species of plants and animals become extinct; diseases die out and fresh ones afflict mankind: all these things doubtless have their causes, but if we do not know what they are, we have no measure of the effects, and cannot tell when or where they will happen.

Laws of Concomitant Variations may hold good only within certain limits. That bodies contract as the temperature falls, is not true of water below 39° F. In Psychology, Weber's Law is only true within the median range of sensation-intensities, not for very faint, nor for very strong, stimuli. In such cases the failure of the laws may depend upon something imperfectly understood in the collocation: as to water, on its molecular constitution; as to sensation, upon the structure of the nervous system.

§ 4. Secondary Laws, again, are either of Succession or of Co-existence.

Those of Succession are either—(1) of direct causation, as that 'Water quenches fire,' or (more strictly) that 'Evaporation reduces temperature'; or (2) of the effect of a remote cause, as 'Bad harvests tend to raise the price of bread'; or (3) of the joint effects of the same cause, as that 'Night follows day' (from the revolution of the earth), or the course of the seasons (from the inclination of the earth's axis).

Laws of Co-existence are of several classes. (1) One has the generality of a primary law, though it is proved only by Agreement, namely, 'All gravitating bodies are inert'. Others, though less general than this, are of very extensive range, as that 'All gases that are not decomposed by rise of temperature have the same rate of expansion'; and, in Botany that 'All monocotyledonous plants are endogenous'. These laws of Co-existence are concerned with fundamental properties of bodies.

(2) Next come laws of the Co-existence of those properties which are comprised in the definitions of Natural Kinds. Mill distinguished between (α) classes of things that agree among themselves and differ from others only in one or a few attributes (such as 'red things,' 'musical notes', 'carnivorous animals', 'soldiers'), and (β) classes of things that agree among themselves and differ from others in a multitude of characters: and the latter he calls Natural Kinds. These comprise the chemical elements and their pure compounds (such as water, alcohol, rock-salt), and the species of plants and animals. Clearly, each of these is constituted by the co-existence or co-inherence of a multitude of properties, some of which are selected as the basis of their definitions. Thus, Gold is a metal of high specific gravity, atomic weight 197.2, high melting point, low chemical affinities, great ductility, yellow colour, *etc.*: a Horse has 'a vertebral column, mammæ, a placental embryo, four legs, a single well-developed toe in each foot provided with a hoof, a bushy tail, and callosities on the inner sides of both the fore and the hind legs' (Huxley).

Since Darwinism has obtained general acceptance, some Logicians have doubted the propriety of calling the organic species 'Kinds,' on the ground that they are not, as to definiteness and permanence, on a par with the chemical elements or such compounds as water and rock-salt; that they vary extensively, and that it is only by the loss of former generations of animals that we are able to distinguish species at all. But to this it may be replied that species are often approximately constant for immense periods of time, and may be called permanent in comparison with human generations; and that, although the leading principles of Logic are perhaps eternal truths, yet upon a detail such as this, the science may condescend to recognise a distinction if it is good for (say) only 100,000 years. That if former generations of plants and animals were not lost, all distinctions of species would disappear, may be true; but they are lost—for the most part beyond hope of recovery; and accordingly the distinction of species is still recognised; although there are cases, chiefly at the lower stages of organisation, in which so many varieties occur as to make adjacent species almost or quite indistinguishable. So far as species are recognised, then, they present a complex co-inherence of qualities, which is, in one aspect, a logical problem; and, in another, a logical datum; and, coming more naturally under the head of Natural Kinds than any other, they must be mentioned in this place.

(3) There are, again, certain coincidences of qualities not essential to any kind, and sometimes prevailing amongst many different kinds: such as 'Insects of nauseous taste have vivid (warning) colours'; 'White tom-cats with blue eyes are deaf'; 'White spots and patches, when they appear in domestic animals, are most frequent on the left side.'

(4) Finally, there may be constancy of relative position, as of sides and angles in Geometry; and also among concrete things (at least for long periods of time), as of the planetary orbits, the apparent positions of fixed stars in the sky, the distribution of land and water on the globe, opposite seasons in opposite hemispheres.

All these cases of Co-existence (except the geometrical) present the problem of deriving them from Causation; for there is no general Law of Co-existence from which they can be derived; and, indeed, if we conceive of the external world as a perpetual redistribution of matter and energy, it follows that the whole state of Nature at any instant, and therefore every co-existence included in it, is due to causation issuing from some earlier distribution of matter and energy. Hence, indeed, it is not likely that the problems of co-existence as a whole will ever be solved, since the original distribution of matter is, of course, unknown. Still, starting with any given state of Nature, we may hope to explain some of the co-existences in any subsequent state. We do not, indeed, know why heavy bodies are always inert, nor why the chemical elements are what they are; but it is known that "the properties of the elements are functions of their atomic weight," which (though, at present, only an empirical law) may be a clue to some deeper explanation. As to plants and animals, we know the conditions of their generation, and can trace a connection between most of their characteristics and the conditions of their life: as that the teeth and stomach of animals vary with their food, and that their colour generally varies with their habitat.

Geometrical Co-existence, when it is not a matter of definition (as 'a square is a rectangle with four equal sides'), is deduced from the definitions and axioms: as when it is shown that in triangles the greater side is opposite the greater angle. The deductions of theorems or secondary laws, in Geometry is a type of what is desirable in the Physical Sciences: the demonstration, namely, that all the connections of phenomena, whether successive or co-existent, are consequences of the redistribution of matter and energy according to the principle of Causation.

Coincidences of Co-existence (Group (3)) may sometimes be deduced and sometimes not. That 'nauseous insects have vivid coloration' comes under the general law of 'protective coloration'; as they are easily recognised and therefore avoided by insectivorous birds and other animals. But why white tom-cats with blue-eyes should be deaf, is (I believe) unknown. When co-existences cannot be derived from causation, they can only be proved by collecting examples and trusting vaguely to the Uniformity of Nature. If no exceptions are found, we have an empirical law of considerable probability within the range of our exploration. If exceptions occur, we have at most an approximate generalisation, as that 'Most metals are whitish,' or 'Most domestic cats are tabbies' (but this probably is the ancestral colouring). We may then resort to statistics for greater definiteness, and find that in Hampshire (say) 90 *per cent.* of the domestic cats are tabby.

§ 5. Scientific Explanation consists in discovering, deducing, and assimilating the laws of phenomena; it is the analysis of that Heracleitan 'flux' which so many philosophers have regarded as intractable to human inquiry. In the ordinary use of the word, 'explanation' means the satisfying a man's understanding; and what may serve this purpose depends partly upon the natural soundness of his understanding, and partly on his education; but it is always at last an appeal to the primary functions of cognition, discrimination and assimilation.

Generally, what we are accustomed to seems to need no explanation, unless our curiosity is particularly directed to it. That boys climb trees and throw stones, and that men go fox-hunting, may easily pass for matters of course. If any one is so exacting as to ask the reason, there is a ready answer in the 'need of exercise.' But this will not explain the peculiar zest of those exercises, which is something quite different from our feelings whilst swinging dumb-bells or tramping the highway. Others, more sophisticated, tell us that the civilised individual retains in his nature the instincts of his remote ancestors, and that these assert themselves at stages of his growth corresponding with ancestral periods of culture or savagery: so that if we delight to climb trees, throw stones, and hunt, it is because our forefathers once lived in trees, had no missiles but stones, and depended for a livelihood upon killing something. To some of us, again, this seems an explanation; to others it merely gives annoyance, as a superfluous hypothesis, the fruit of a wanton imagination and too much leisure.

However, what we are not accustomed to immediately excites curiosity. If it were exceptional to climb trees, throw stones, ride after foxes, whoever did such things would be viewed with suspicion. An eclipse, a shooting star, a solitary boulder on the heath, a strange animal, or a Chinaman in the street, calls for explanation; and among some nations, eclipses have been explained by supposing a dragon to devour the sun or moon; solitary boulders, as the missiles of a giant; and so on. Such explanations, plainly, are attempts to regard rare phenomena as similar to others that are better known; a snake having been seen to swallow a rabbit, a bigger one may swallow the sun: a giant is supposed to bear much the same relation to a boulder as a boy does to half a brick. When any very common thing seems to need no explanation, it is because the several instances of its occurrence are a sufficient basis of assimilation to satisfy most of us. Still, if a reason for such a thing be demanded, the commonest answer has the same implication, namely, that assimilation or classification is a sufficient reason for it. Thus, if climbing trees is referred to the need of exercise, it is assimilated to running, rowing, *etc.*; if the customs of a savage tribe are referred to the command of its gods, they are assimilated to those things that are done at the command of chieftains.

Explanation, then, is a kind of classification; it is the finding of resemblance between the phenomenon in question and other phenomena. In Mathematics, the explanation of a theorem is the same as its proof, and consists in showing that it repeats, under different conditions, the definitions and axioms already assumed and the theorems already demonstrated. In Logic, the major premise of every syllogism is an explanation of the conclusion; for the minor premise asserts that the conclusion is an example of the major premise.

In Concrete Sciences, to discover the cause of a phenomenon, or to derive an empirical law from laws of causation, is to explain it; because a cause is an invariable antecedent, and therefore reminds us of, or enables us to conceive, an indefinite number of cases similar to the present one wherever the cause exists. It classifies the present case with other instances of causation, or brings it under the universal law; and, as we have seen that the discovery of the laws of nature is essentially the discovery of causes, the discovery and derivation of laws is scientific explanation.

The discovery of quantitative laws is especially satisfactory, because it not only explains why an event happens at all, but why it happens just in this direction, degree, or amount; and not only is the given relation of cause and effect definitely assimilated to other causal instances, but the effect is identified with the cause as the same matter and energy redistributed; wherefore, whether the conservation of matter and energy be universally true or not, it must still be an universal postulate of scientific explanation.

The mere discovery of an empirical law of co-existence, as that 'white tom-cats with blue eyes are deaf', is indeed something better than an isolated fact: every general proposition relieves the mind of a load of facts; and, for many people, to be able to say—'It is always so'—may be enough; but for scientific explanation we require to know the reason of it, that is, the cause. Still, if asked to explain an axiom, we can only say, 'It is always so:' though it is some relief to point out particular instances of its realisation, or to exhibit the similarity of its form to that of other axioms—as of the *Dictum* to the axiom of equality.

§ 6. There are three modes of scientific Explanation; First, the analysis of a phenomenon into the laws of its causes and the concurrence of those causes.

The pumping of water implies (1) pressure of the air, (2) distribution of pressure in a liquid, (3) that motion takes the direction of least resistance. Similarly, that thunder follows forked lightning, and that the report of a gun follows the flash, are resolvable into (1) the discharge of electricity, or the explosion of gunpowder; (2) distance of the observer from the event; (3) that light travels faster than sound. The planetary orbits are analysable into the tendency of planets to fall into the sun, and their tendency to travel in a straight line. When this conception is helped out by swinging a ball round by a string, and then letting it go, to show what would happen to the earth if gravitation ceased, we see how the recognition of resemblance lies at the bottom of explanation.

Secondly, the discovery of steps of causation between a cause and its remote effects; the interpolation and concatenation of causes.

The maxim 'No cats no clover' is explained by assigning the intermediate steps in the following series; that the fructification of red clover depends on the visits of humble-bees, who distribute the pollen in seeking honey; that if field-mice are numerous they destroy the humble-bees' nests; and that (owls and weasels being exterminated by gamekeepers) the destruction of field-mice depends upon the supply of cats; which, therefore, are a remote condition of the clover crop. Again, the communication of thought by speech is an example of something so common that it seems to need no explanation; yet to explain it is a long story. A thought in one man's mind is the remote cause of a similar thought in another's: here we have (1) a thought associated with mental words; (2) a connection between these thoughts and some tracts of the brain; (3) a connection between these tracts of the brain and the muscles of the larynx, the tongue and the lips; (4) movements of the chest, larynx and mouth, propelling and modifying waves of air; (5) the impinging of these air-waves upon another man's ear, and by a complex mechanism exciting the aural nerve; (6) the transfer of this excitation to certain tracts of his brain; (7) a connection there with sounds of words and their associated thoughts. If one of these links fail, there is no communication.

Thirdly, the subsumption of several laws under one more general expression.

The tendency of bodies to fall to the earth and the tendency of the earth itself (with the other planets) to fall into the sun, are subsumed under the general law that 'All matter gravitates.' The same law subsumes the movements of the tide. By means of the notion of specific gravity, it includes 'levitation,' or the actual rising of some bodies, as of corks in water, of balloons, or flames in the air: the fact being that these things do not tend to rise, but to fall like everything else; only as the water or air weighs more in proportion to its volume than corks or balloons, the latter are pushed up.

This process of subsumption bears the same relation to secondary laws, that these do to particular facts. The generalisation of many particular facts (that is, a statement of that in which they agree) is a law; and the generalisation of these laws (that is, again, a statement of that in which they agree) is a higher law; and this process, upwards or downwards, is characteristic of scientific progress. The perfecting of any science consists in comprehending more and more of the facts within its province, and in showing that they all exemplify a smaller and smaller number of principles, which express their most profound resemblances.

These three modes of explanation (analysis, interpolation, subsumption) all consist in generalising or assimilating the phenomena. The pressure of the air, of a liquid, and motion in the direction of least resistance, are all commoner facts than pumping; that light travels faster than sound is a commoner fact than a thunderstorm or gun-firing. Each of the laws—'Cats kill mice,' 'Mice destroy humble-bees' nests,' 'Humble-bees fructify red clover'—is wider and expresses the resemblance of more numerous cases than the law that 'Clover depends on cats'; because each of them is less subject to further conditions. Similarly, every step in the communication of thought by language is less conditional, and therefore more general, than the completion of the process.

In all the above cases, again, each law into which the phenomenon (whether pumping or conversation) is resolved, suggests a host of parallel cases: as the modifying of air-waves by the larynx and lips suggests the various devices by which the strings and orifices of musical instruments modify the character of notes.

Subsumption consists entirely in proving the existence of an essential similarity between things where it was formerly not observed: as that the gyrations of the moon, the fall of apples, and the flotation of bubbles are all examples of gravitation: or that the purifying of the blood by breathing, the burning of a candle, and the rusting of iron are all cases of oxidation: or that the colouring of the underside of a red-admiral's wings, the spots of the giraffe, the shape and attitude of a stick-caterpillar, the immobility of a bird on its nest, and countless other cases, though superficially so different, agree in this, that they conceal and thereby protect the organism.

Not any sort of likeness, however, suffices for scientific explanation: the only satisfactory explanation of concrete things or events, is to discover their likeness to others in respect of Causation. Hence attempts to help the understanding by familiar comparisons are often worse than useless. Any of the above examples will show that the first result of explanation is not to make a phenomenon seem familiar, but to put (as the saying is) 'quite a new face upon it.' When, indeed, we have thought it over in all its newly discovered relations, we feel more at home with it than ever; and this is one source of our satisfaction in explaining things; and hence to substitute immediate familiarisation for radical explanation, is the easily besetting sin of human understanding: the most plausible of fallacies, the most attractive, the most difficult to avoid even when we are on our guard against it.

§ 7. The explanation of Nature (if it be admitted to consist in generalisation, or the discovery of resemblance amidst differences) can never be completed. For—(1) there are (as Mill says) facts, namely, fundamental states or processes of consciousness, which are distinct; in other words, they do not resemble one another, and therefore cannot be generalised or subsumed under one explanation. Colour, heat, smell, sound, touch, pleasure and pain, are so different that there is one group of conditions to be sought for each; and the laws of these conditions cannot be subsumed under a more general one without leaving out the very facts to be explained. A general condition of sensation, such as the stimulating of the sensory organs of a living animal, gives no account of the *special* characters of colour, smell, *etc.*; which are, however, the phenomena in question; and each of them has its own law. Nay, each distinct sensation-quality, or degree, must have its own law; for in each ultimate difference there is something that cannot be assimilated. Such differences amount, according to experimental Psychologists, to more than 50,000. Moreover, a neural process can never explain a conscious process in the way of cause and effect; for there is no equivalence between them, and one can never absorb the other.

(2) When physical science is treated objectively (that is, with as little reference as possible to the fact that all phenomena are only known in relation to the human mind), colour, heat, smell, sound (considered as sensations) are neglected, and attention is fixed upon certain of their conditions: extension, figure, resistance, weight, motion, with their derivatives, density, elasticity, *etc.* These are called the Primary Qualities of Matter; and it is assumed that they belong to matter by itself, whether we look on or not: whilst colour, heat, sound, *etc.*, are called Secondary Qualities, as depending entirely upon the reaction of some conscious animal. By physical science the world is considered in the abstract, as a perpetual redistribution of matter and energy, and the distracting multiplicity of sensations seems to be got rid of.

But, not to dwell upon the difficulty of reducing the activities of life and chemistry to mechanical principles—even if this were done, complete explanation could not be attained. For—(*a*) as explanation is the discovery of causes, we no sooner succeed in assigning the causes of the present state of the world than we have to inquire into the causes of those causes, and again the still earlier causes, and so on to infinity. But, this being impossible, we must be content, wherever we stop, to contemplate the uncaused, that is, the unexplained; and then all that follows is only relatively explained.

Besides this difficulty, however, there is another that prevents the perfecting of any theory of the abstract material world, namely (*b*), that it involves more than one first principle. For we have seen that the Uniformity of Nature is not really a principle, but a merely nominal generalisation, since it cannot be definitely stated; and, therefore, the principles of Contradiction, Mediate Equality, and Causation remain incapable of subsumption; nor can any one of them be reduced to another: so that they remain unexplained.

(3) Another limit to explanation lies in the infinite character of every particular fact; so that we may know the laws of many of its properties and yet come far short of understanding it as a whole. A lump of sandstone in the road: we may know a good deal about its specific gravity, temperature, chemical composition, geological conditions; but if we inquire the causes of the particular modifications it exhibits of these properties, and further why it is just so big, containing so many molecules, neither more nor less, disposed in just such relations to one another as to give it this particular figure, why it lies exactly there rather than a yard off, and so forth, we shall get no explanation of all this. The causes determining each particular phenomenon are infinite, and can never be computed; and, therefore, it can never be fully explained.

§ 8. Analogy is used in two senses: (1) for the resemblance of relations between terms that have little or no resemblance—as *The wind drives the clouds as a shepherd drives his sheep*—where wind and shepherd, clouds and sheep are totally unlike. Such analogies are a favourite figure in poetry and rhetoric, but cannot prove anything. For valid reasoning there must be parallel cases, according to substance and attribute, or cause and effect, or proportion: *e.g. As cattle and deer are to herbivorousness, so are camels; As bodies near the earth fall toward it, so does the moon; As 2 is to 3 so is 4 to 6.*

(2) Analogy is discussed in Logic as a kind of probable proof based upon imperfect similarity (as the best that can be discovered) between the *data* of comparison and the subject of our inference. Like Deduction and Induction, it assumes that things which are alike in some respects are also alike in others; but it differs from them in not appealing to a definite general law assigning the essential points of resemblance upon which the argument relies. In Deductive proof, this is done by the major premise of every syllogism: if the major says that 'All fat men are humorists,' and we can establish the minor, 'X is a fat man,' we have secured the essential resemblance that carries the conclusion. In induction, the Law of Causation and its representatives, the Canons, serve the same purpose, specifying the essential marks of a cause. But, in Analogy, the resemblance relied on cannot be stated categorically.

If we argue that Mars is inhabited because it resembles the datum, our Earth, (1) in being a planet, (2) neither too hot nor too cold for life, (3) having an atmosphere, (4) land and water, *etc.*, we are not prepared to say that 'All planets having these characteristics are inhabited.' It is, therefore, not a deduction; and since we do not know the original causes of life on the Earth, we certainly cannot show by induction that adequate causes exist in Mars. We rely, then, upon some such vague notion of Uniformity as that 'Things alike in some points are alike in others'; which, plainly, is either false or nugatory. But if the linear markings upon the surface of Mars indicate a system of canals, the inference that he has intelligent inhabitants is no longer analogical, since canals can have no other cause.

The cogency of any proof depends upon the *character* and *definiteness* of the likeness which one phenomenon bears to another; but Analogy trusts to the general *quantity* of likeness between them, in ignorance of what may be the really important likeness. If, having tried with a stone, an apple, a bullet, *etc.*, we find that they all break an ordinary window, and thence infer that a cricket ball will do so, we do not reason by analogy, but make instinctively a deductive extension of an induction, merely omitting the explicit generalisation, 'All missiles of a certain weight, size and solidity break windows.' But if, knowing nothing of snakes except that the viper is venomous, a child runs away from a grass-snake, he argues by analogy; and, though his conduct is prudentially justifiable, his inference is wrong: for there is no law that 'All snakes are venomous,' but only that those are venomous that have a certain structure of fang; a point which he did not stay to examine.

The discovery of an analogy, then, may suggest hypotheses; it states a problem—to find the causes of the analogy; and thus it may lead to scientific proof; but merely analogical argument is only probable in various degrees. (1) The greater the number and importance of the points of agreement, the more probable is the inference. (2) The greater the number and importance of the points of difference, the less probable is the inference. (3) The greater the number of unknown properties in the subject of our argument, the less the value of any inference from those that we do know. Of course the number of unknown properties can itself be estimated only by analogy. In the case of Mars, they are probably very numerous; and, apart from the evidence of canals, the prevalent assumption that there are intelligent beings in that planet, seems to rest less upon probability than on a curiously imaginative extension of the gregarious sentiment, the chilly discomfort of mankind at the thought of being alone in the universe, and a hope that there may be conversable and 'clubable' souls nearer than the Dog-star.

CHAPTER XX
PROBABILITY

§ 1. Chance was once believed to be a distinct power in the world, disturbing the regularity of Nature; though, according to Aristotle, it was only operative in occurrences below the sphere of the moon. As, however, it is now admitted that every event in the world is due to some cause, if we can only trace the connection, whilst nevertheless the notion of Chance is still useful when rightly conceived, we have to find some other ground for it than that of a spontaneous capricious force inherent in things. For such a conception can have no place in any logical interpretation of Nature: it can never be inferred from a principle, seeing that every principle expresses an uniformity; nor, again, if the existence of a capricious power be granted, can any inference be drawn from it. Impossible alike as premise and as conclusion, for Reason it is nothing at all.

Every event is a result of causes: but the multitude of forces and the variety of collocations being immeasurably great, the overwhelming majority of events occurring about the same time are only related by Causation so remotely that the connection cannot be followed. Whilst my pen moves along the paper, a cab rattles down the street, bells in the neighbouring steeple chime the quarter, a girl in the next house is practising her scales, and throughout the world innumerable events are happening which may never happen together again; so that should one of them recur, we have no reason to expect any of the others. This is Chance, or chance coincidence. The word Coincidence is vulgarly used only for the inexplicable concurrence of *interesting* events—"quite a coincidence!"

On the other hand, many things are now happening together or coinciding, that will do so, for assignable reasons, again and again; thousands of men are leaving the City, who leave at the same hour five days a week. But this is not chance; it is causal coincidence due to the custom of business in this country, as determined by our latitude and longitude and other circumstances. No doubt the above chance coincidences—writing, cab-rattling, chimes, scales, *etc.*—are causally connected at some point of past time. They were predetermined by the condition of the world ten minutes ago; and that was due to earlier conditions, one behind the other, even to the formation of the planet. But whatever connection there may have been, we have no such knowledge of it as to be able to deduce the coincidence, or calculate its recurrence. Hence Chance is defined by Mill to be: Coincidence giving no ground to infer uniformity.

Still, some chance coincidences do recur according to laws of their own: I say *some*, but it may be all. If the world is finite, the possible combinations of its elements are exhaustible; and, in time, whatever conditions of the world have concurred will concur again, and in the same relation to former conditions. This writing, that cab, those chimes, those scales will coincide again; the Argonautic expedition, and the Trojan war, and all our other troubles will be renewed. But let us consider some more manageable instance, such as the throwing of dice. Every one who has played much with dice knows that double sixes are sometimes thrown, and sometimes double aces. Such coincidences do not happen once and only once; they occur again and again, and a great number of trials will show that, though their recurrence has not the regularity of cause and effect, it yet has a law of its own, namely—a tendency to average regularity. In 10,000 throws there will be some number of double sixes; and the greater the number of throws the more closely will the average recurrence of double sixes, or double aces, approximate to one in thirty-six. Such a law of average recurrence is the basis of Probability. Chance being the fact of coincidence without assignable cause, Probability is expectation based on the average frequency of its happening.

§ 2. Probability is an ambiguous term. Usually, when we say that an event is 'probable,' we mean that it is more likely than not to happen. But, scientifically, an event is probable if our expectation of its occurrence is less than certainty, as long as the event is not impossible. Probability, thus conceived, is represented by a fraction. Taking 1 to stand for certainty, and 0 for impossibility, probability may be $999/1000$, or $1/1000$, or (generally) $1/m$. The denominator represents the number of times that an event happens, and the numerator the number of times that it coincides with another event. In throwing a die, the probability of ace turning up is expressed by putting the number of throws for the denominator and the number of times that ace is thrown for the numerator; and we may assume that the more trials we make the nearer will the resulting fraction approximate to $1/6$.

Instead of speaking of the 'throwing of the die' and its 'turning up ace' as two events, the former is called 'the event' and the latter 'the way of its happening.' And these expressions may easily be extended to cover relations of distinct events; as when two men shoot at a mark and we desire to represent the probability of both hitting the bull's eye together, each shot may count as an event (denominator) and the coincidence of 'bull's-eyes' as the way of its happening (numerator).

It is also common to speak of probability as a proportion. If the fraction expressing the probability of ace being cast is 1/6, the proportion of cases in which it happens is 1 to 5; or (as it is, perhaps, still more commonly put) 'the chances are 5 to 1 against it.'

§ 3. As to the grounds of probability opinions differ. According to one view the ground is subjective: probability depends, it is said, upon the quantity of our Belief in the happening of a certain event, or in its happening in a particular way. According to the other view the ground is objective, and, in fact, is nothing else than experience, which is most trustworthy when carefully expressed in statistics.

To the subjective view it may be objected, (*a*) that belief cannot by itself be satisfactorily measured. No one will maintain that belief, merely as a state of mind, always has a definite numerical value of which one is conscious, as 1/100 or 1/10. Let anybody mix a number of letters in a bag, knowing nothing of them except that one of them is X, and then draw them one by one, endeavouring each time to estimate the value of his belief that the next will be X; can he say that his belief in the drawing of X next time regularly increases as the number of letters left decreases?

If not, we see that (*b*) belief does not uniformly correspond with the state of the facts. If in such a trial as proposed above, we really wish to draw X, as when looking for something in a number of boxes, how common it is, after a few failures, to feel quite hopeless and to say: "Oh, of course it will be in the last." For belief is subject to hope and fear, temperament, passion, and prejudice, and not merely to rational considerations. And it is useless to appeal to 'the Wise Man,' the purely rational judge of probability, unless he is producible. Or, if it be said that belief is a short cut to the evaluation of experience, because it is the resultant of all past experience, we may reply that this is not true. For one striking experience, or two or three recent ones, will immensely outweigh a great number of faint or remote experiences. Moreover, the experience of two men may be practically equal, whilst their beliefs upon any question greatly differ. Any two Englishmen have about the same experience, personal and ancestral, of the weather; yet their beliefs in the saw that 'if it rain on St. Swithin's Day it will rain for forty days after,' may differ as confident expectation and sheer scepticism. Upon which of these beliefs shall we ground the probability of forty days' rain?

But (*c*) at any rate, if Probability is to be connected with Inductive Logic, it must rest upon the same ground, namely—observation. Induction, in any particular case, is not content with beliefs or opinions, but aims at testing, verifying or correcting them by appealing to the facts; and Probability has the same object and the same basis.

In some cases, indeed, the conditions of an event are supposed to be mathematically predetermined, as in tossing a penny, throwing dice, dealing cards. In throwing a die, the ways of happening are six; in tossing a penny only two, head and tail: and we usually assume that the odds with a die are fairly 5 to 1 against ace, whilst with a penny 'the betting is even' on head or tail. Still, this assumption rests upon another, that the die is perfectly fair, or that the head and tail of a penny are exactly alike; and this is not true. With an ordinary die or penny, a very great number of trials would, no doubt, give an average approximating to 1/6 or 1/2; yet might always leave a certain excess one way or the other, which would also become more definite as the trials went on; thus showing that the die or penny did not satisfy the mathematical hypothesis. Buffon is said to have tossed a coin 4040 times, obtaining 1992 heads and 2048 tails; a pupil of De Morgan tossed 4092 times, obtaining 2048 heads and 2044 tails.

There are other important cases in which probability is estimated and numerically expressed, although statistical evidence directly bearing upon the point in question cannot be obtained; as in betting upon a race; or in the prices of stocks and shares, which are supposed to represent the probability of their paying, or continuing to pay, a certain rate of interest. But the judgment of experts in such matters is certainly based upon experience; and great pains are taken to make the evidence as definite as possible by comparing records of speed, or by financial estimates; though something must still be allowed for reports of the condition of horses, or of the prospects of war, harvests, *etc.*

However, where statistical evidence is obtainable, no one dreams of estimating probability by the quantity of his belief. Insurance offices, dealing with fire, shipwreck, death, accident, *etc.*, prepare elaborate statistics of these events, and regulate their rates accordingly. Apart from statistics, at what rate ought the lives of men aged 40 to be insured, in order to leave a profit of 5 per cent. upon £1000 payable at each man's death? Is 'quantity of belief' a sufficient basis for doing this sum?

§ 4. The ground of probability is experience, then, and, whenever possible, statistics; which are a kind of induction. It has indeed been urged that induction is itself based upon probability; that the subtlety, complexity and secrecy of nature are such, that we are never quite sure that we fully know even what we have observed; and that, as for laws, the conditions of the universe at large may at any moment be completely changed; so that all imperfect inductions, including the law of causation itself, are only probable. But, clearly, this doctrine turns upon another ambiguity in the word 'probable.' It may be used in the sense of 'less than absolutely certain'; and such doubtless is the condition of all human knowledge, in comparison with the comprehensive intuition of arch-angels: or it may mean 'less than certain according to *our* standard of certainty,' that is, in comparison with the law of causation and its derivatives.

We may suppose some one to object that "by this relative standard even empirical laws cannot be called 'only probable' as long as we 'know no exception to them'; for that is all that can be said for the boasted law of causation; and that, accordingly, we can frame no fraction to represent their probability. That 'all swans are white' was at one time, from this point of view, not probable but certain; though we now know it to be false. It would have been an indecorum to call it only probable as long as no other-coloured swan had been discovered; not merely because the quantity of belief amounted to certainty, but because the number of events (seeing a swan) and the number of their happenings in a certain way (being white) were equal, and therefore the evidence amounted to 1 or certainty." But, in fact, such an empirical law is only probable; and the estimate of its probability must be based on the number of times that similar laws have been found liable to exceptions. Albinism is of frequent occurrence; and it is common to find closely allied varieties of animals differing in colour. Had the evidence been duly weighed, it could never have seemed more than probable that 'all swans are white.' But what law, approaching the comprehensiveness of the law of causation, presents any exceptions?

Supposing evidence to be ultimately nothing but accumulated experience, the amount of it in favour of causation is incomparably greater than the most that has ever been advanced to show the probability of any other kind of event; and every relation of events which is shown to have the marks of causation obtains the support of that incomparably greater body of evidence. Hence the only way in which causation can be called probable, for us, is by considering it as the upward limit (1) to which the series of probabilities tends; as impossibility is the downward limit (0). Induction, 'humanly speaking,' does not rest on probability; but the probability of concrete events (not of mere mathematical abstractions like the falling of absolutely true dice) rests on induction and, therefore, on causation. The inductive evidence underlying an estimate of probability may be of three kinds: (*a*) direct statistics of the events in question; as when we find that, at the age of 20, the average expectation of life is 39-40 years. This is an empirical law, and, if we do not know the causes of any event, we must be content with an empirical law. But (*b*) if we do know the causes of an event, and the causes which may prevent its happening, and can estimate the comparative frequency of their occurring, we may deduce the probability that the effect (that is, the event in question) will occur. Or (*c*) we may combine these two methods, verifying each by means of the other. Now either the method (*b*) or (*a fortiori*) the method (*c*) (both depending on causation) is more trustworthy than the method (*a*) by itself.

But, further, a merely empirical statistical law will only be true as long as the causes influencing the event remain the same. A die may be found to turn ace once in six throws, on the average, in close accordance with mathematical theory; but if we load it on that facet the results will be very different. So it is with the expectation of life, or fire, or shipwreck. The increased virulence of some epidemic such as influenza, an outbreak of anarchic incendiarism, a moral epidemic of over-loading ships, may deceive the hopes of insurance offices. Hence we see, again, that probability depends upon causation, not causation upon probability.

That uncertainty of an event which arises not from ignorance of the law of its cause, but from our not knowing whether the cause itself does or does not occur at any particular time, is Contingency.

§ 5. The nature of an average supposes deviations from it. Deviations from an average, or "errors," are assumed to conform to the law (1) that the greater errors are less frequent than the smaller, so that most events approximate to the average; and (2) that errors have no "bias," but are equally frequent and equally great in both directions from the mean, so that they are scattered symmetrically. Hence their distribution may be expressed by some such figure as the following:

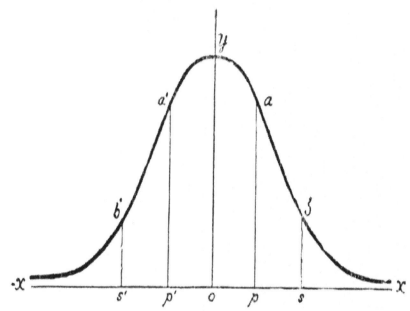

Fig. 11.

Here *o* is the average event, and *oy* represents the number of average events. Along *ox*, in either direction, deviations are measured. At *p* the amount of error or deviation is *op*; and the number of such deviations is represented by the line or ordinate *pa*. At *s* the deviation is *os*; and the number of such deviations is expressed by *sb*. As the deviations grow greater, the number of them grows less. On the other side of *o*, toward -*x*, at distances, *op'*, *os'* (equal to *op*, *os*) the lines *p'a'*, *s'b'* represent the numbers of those errors (equal to *pa*, *sb*).

If *o* is the average height of the adult men of a nation, (say) 5 ft. 6 in., *s'* and *s* may stand for 5 ft. and 6 ft.; men of 4 ft. 6 in. lie further toward -*x*, and men of 6 ft. 6 in. further toward *x*. There are limits to the stature of human beings (or to any kind of animal or plant) in both directions, because of the physical conditions of generation and birth. With such events the curve *b'yb* meets the abscissa at some point in each direction; though where this occurs can only be known by continually measuring dwarfs and giants. But in throwing dice or tossing coins, whilst the average occurrence of ace is once in six throws, and the average occurrence of 'tail' is once in two tosses, there is no necessary limit to the sequences of ace or of 'tail' that may occur in an infinite number of trials. To provide for such cases the curve is drawn as if it never touched the abscissa.

That some such figure as that given above describes a frequent characteristic of an average with the deviations from it, may be shown in two ways: (1) By arranging the statistical results of any homogeneous class of measurements; when it is often found that they do, in fact, approximately conform to the figure; that very many events are near the average; that errors are symmetrically distributed on either side, and that the greater errors are the rarer. (2) By mathematical demonstration based upon the supposition that each of the events in question is influenced, more or less, by a number of unknown conditions common to them all, and that these conditions are independent of one another. For then, in rare cases, all the conditions will operate favourably in one way, and the men will be tall; or in the opposite way, and the men will be short; in more numerous cases, many of the conditions will operate in one direction, and will be partially cancelled by a few opposing them; whilst in still more cases opposed conditions will approximately balance one another and produce the average event or something near it. The results will then conform to the above figure.

From the above assumption it follows that the symmetrical curve describes only a 'homogeneous class' of measurements; that is, a class no portion of which is much influenced by conditions peculiar to itself. If the class is not homogeneous, because some portion of it is subject to *peculiar* conditions, the curve will show a hump on one side or the other. Suppose we are tabulating the ages at which Englishmen die who have reached the age of 20, we may find that the greatest number die at 39 (19 years being the average expectation of life at 20) and that as far as that age the curve upwards is regular, and that beyond the age of 39 it begins to descend regularly, but that on approaching 45 it bulges out some way before resuming its regular descent—thus:

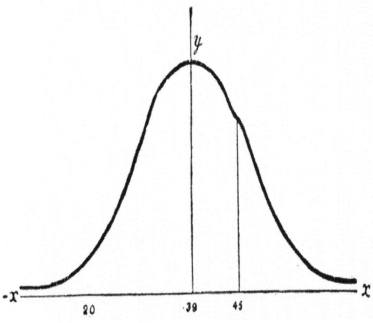

Fig. 12.

Such a hump in the curve might be due to the presence of a considerable body of teetotalers, whose longevity was increased by the peculiar condition of abstaining from alcohol, and whose average age was 45, 6 years more than the average for common men.

Again, if the group we are measuring be subject to selection (such as British soldiers, for which profession all volunteers below a certain height—say, 5 ft. 5 in.—are rejected), the curve will fall steeply on one side, thus:

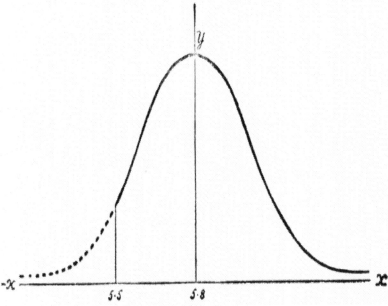

Fig. 13.

If, above a certain height, volunteers are also rejected, the curve will fall abruptly on both sides. The average is supposed to be 5 ft. 8 in.

The distribution of events is described by 'some such curve' as that given in Fig. 11; but different groups of events may present figures or surfaces in which the slopes of the curves are very different, namely, more or less steep; and if the curve is very steep, the figure runs into a peak; whereas, if the curve is gradual, the figure is comparatively flat. In the latter case, where the figure is flat, fewer events will closely cluster about the average, and the deviations will be greater.

Suppose that we know nothing of a given event except that it belongs to a certain class or series, what can we venture to infer of it from our knowledge of the series? Let the event be the cephalic index of an Englishman. The cephalic index is the breadth of a skull × 100 and divided by the length of it; *e.g.* if a skull is 8 in. long and 6 in. broad, (6×100)/8=75. We know that the average English skull has an index of 78. The skull of the given individual, therefore, is more likely to have that index than any other. Still, many skulls deviate from the average, and we should like to know what is the probable error in this case. The probable error is the measurement that divides the deviations from the average in either direction into halves, so that there are as many events having a greater deviation as there are events having a less deviation. If, in Fig. 11 above, we have arranged the measurements of the cephalic index of English adult males, and if at o (the average or mean) the index is 78, and if the line pa divides the right side of the fig. into halves, then op is the probable error. If the measurement at p is 80, the probable error is 2. Similarly, on the left hand, the probable error is op', and the measurement at p' is 76. We may infer, then, that the skull of the man before us is more likely to have an index of 78 than any other; if any other, it is equally likely to lie between 80 and 76, or to lie outside them; but as the numbers rise above 80 to the right, or fall below 76 to the left, it rapidly becomes less and less likely that they describe this skull.

In such cases as heights of men or skull measurements, where great numbers of specimens exist, the average will be actually presented by many of them; but if we take a small group, such as the measurements of a college class, it may happen that the average height (say, 5 ft. 8 in.) is not the actual height of any one man. Even then there will generally be a closer cluster of the actual heights about that number than about any other. Still, with very few cases before us, it may be better to take the median than the average. The median is that event on either side of which there are equal numbers of deviations. One advantage of this procedure is that it may save time and trouble. To find approximately the average height of a class, arrange the men in order of height, take the middle one and measure him. A further advantage of this method is that it excludes the influence of extraordinary deviations. Suppose we have seven cephalic indices, from skeletons found in the same barrow, 75½, 76, 78, 78, 79, 80½, 86. The average is 79; but this number is swollen unduly by the last measurement; and the median, 78, is more fairly representative of the series; that is to say, with a greater number of skulls the average would probably have been nearer 78.

To make a single measurement of a phenomenon does not give one much confidence. Another measurement is made; and then, if there is no opportunity for more, one takes the mean or average of the two. But why? For the result may certainly be worse than the first measurement. Suppose that the events I am measuring are in fact fairly described by Fig. II, although (at the outset) I know nothing about them; and that my first measurement gives p, and my second s; the average of them is worse than p. Still, being yet ignorant of the distribution of these events, I do rightly in taking the average. For, as it happens, ¾ of the events lie to the left of p; so that if the first trial gives p, then the average of p and any subsequent trial that fell nearer than (say) s' on the opposite side, would be better than p; and since deviations greater than s' are rare, the chances are nearly 3 to 1 that the taking of an average will improve the observation. Only if the first trial give o, or fall within a little more than ½p on either side of o, will the chances be against any improvement by trying again and taking an average. Since, therefore, we cannot know the position of our first trial in relation to o, it is always prudent to try again and take the average; and the more trials we can make and average, the better is the result. The average of a number of observations is called a "Reduced Observation."

We may have reason to believe that some of our measurements are better than others because they have been taken by a better trained observer, or by the same observer in a more deliberate way, or with better instruments, and so forth. If so, such observations should be 'weighted,' or given more importance in our calculations; and a simple way of doing this is to count them twice or oftener in taking the average.

§ 6. These considerations have an important bearing upon the interpretation of probabilities. The average probability for any *general class* or series of events cannot be confidently applied to any *one instance* or to any *special class* of instances, since this one, or this special class, may exhibit a striking error or deviation; it may, in fact, be subject to special causes. Within the class whose average is first taken, and which is described by general characters as 'a man,' or 'a die,' or 'a rifle shot,' there may be classes marked by special characters and determined by special influences. Statistics giving the average for 'mankind' may not be true of 'civilised men,' or of any still smaller class such as 'Frenchmen.' Hence life-insurance offices rely not merely on statistics of life and death in general, but collect special evidence in respect of different ages and sexes, and make further allowance for teetotalism, inherited disease, *etc*. Similarly with individual cases: the average expectation for a class, whether general or special, is only applicable to any particular case if that case is adequately described by the class characters. In England, for example, the average expectation of life for males at 20 years of age is 39.40; but at 60 it is still 13.14, and at 73 it is 7.07; at 100 it's 1.61. Of men 20 years old those who live more or less than 39.40 years are deviations or errors; but there are a great many of them. To insure the life of a single man at 20, in the expectation of his dying at 60, would be a mere bet, if we had no special knowledge of him; the safety of an insurance office lies in having so many clients that opposite deviations cancel one another: the more clients the safer the business. It is quite possible that a hundred men aged 20 should be insured in one week and all of them die before 25; this would be ruinous, if others did not live to be 80 or 90.

Not only in such a practical affair as insurance, but in matters purely scientific, the minute and subtle peculiarities of individuals have important consequences. Each man has a certain cast of mind, character, physique, giving a distinctive turn to all his actions even when he tries to be normal. In every employment this determines his Personal Equation, or average deviation from the normal. The term Personal Equation is used chiefly in connection with scientific observation, as in Astronomy. Each observer is liable to be a little wrong, and this error has to be allowed for and his observations corrected accordingly.

The use of the term 'expectation,' and of examples drawn from insurance and gambling, may convey the notion that probability relates entirely to future events; but if based on laws and causes, it can have no reference to point of time. As long as conditions are the same, events will be the same, whether we consider uniformities or averages. We may therefore draw probable inferences concerning the past as well as the future, subject to the same hypothesis, that the causes affecting the events in question were the same and similarly combined. On the other hand, if we know that conditions bearing on the subject of investigation, have changed since statistics were collected, or were different at some time previous to the collection of evidence, every probable inference based on those statistics must be corrected by allowing for the altered conditions, whether we desire to reason forwards or backwards in time.

§ 7. The rules for the combination of probabilities are as follows:

Little Women

(1) If two events or causes do not concur, the probability of one or the other occurring is the sum of the separate probabilities. A die cannot turn up both ace and six; but the probability in favour of each is 1/6: therefore, the probability in favour of one or the other is 1/3. Death can hardly occur from both burning and drowning: if 1 in 1000 is burned and 2 in 1000 are drowned, the probability of being burned or drowned is 3/1000.

(2) If two events are independent, having neither connection nor repugnance, the probability of their concurring is found by multiplying together the separate probabilities of each occurring. If in walking down a certain street I meet A once in four times, and B once in three times, I ought (by mere chance) to meet both once in twelve times: for in twelve occasions I meet B four times; but once in four I meet A.

This is a very important rule in scientific investigation, since it enables us to detect the presence of causation. For if the coincidence of two events is more or less frequent than it would be if they were entirely independent, there is either connection or repugnance between them. If, *e.g.*, in walking down the street I meet both A and B oftener than once in twelve times, they may be engaged in similar business, calling them from their offices at about the same hour. If I meet them both less often than once in twelve times, they may belong to the same office, where one acts as a substitute for the other. Similarly, if in a multitude of throws a die turns six oftener than once in six times, it is not a fair one: that is, there is a cause favouring the turning of six. If of 20,000 people 500 see apparitions and 100 have friends murdered, the chance of any man having both experiences is 1/8000; but if each lives on the average 300,000 hours, the chance of both events occurring in the same hour is 1/2400000000. If the two events occur in the same hour oftener than this, there is more than a chance coincidence.

The more minute a cause of connection or repugnance between events, the longer the series of trials or instances necessary to bring out its influence: the less a die is loaded, the more casts must be made before it can be shown that a certain side tends to recur oftener than once in six.

(3) The rule for calculating the probability of a dependent event is the same as the above; for the concurrence of two independent events is itself dependent upon each of them occurring. My meeting with both A and B in the street is dependent on my walking there and on my meeting one of them. Similarly, if A is sometimes a cause of B (though liable to be frustrated), and B sometimes of C (C and B having no causes independent of B and A respectively), the occurrence of C is dependent on that of B, and that again on the occurrence of A. Hence we may state the rule: If two events are dependent each on another, so that if one occur the second may (or may not), and if the second a third; whilst the third never occurs without the second, nor the second without the first; the probability that if the first occur the third will, is found by multiplying together the fractions expressing the probability that the first is a mark of the second and the second of the third.

Upon this principle the value of hearsay evidence or tradition deteriorates, and generally the cogency of any argument based upon the combination of approximate generalisations dependent on one another or "self-infirmative." If there are two witnesses, A and B, of whom A saw an event, whilst B only heard A relate it (and is therefore dependent on A), what credit is due to B's recital? Suppose the probability of each man's being correct as to what he says he saw, or heard, is 3/4: then ($3/4 \times 3/4 = 9/16$) the probability that B's story is true is a little more than 1/2. For if in 16 attestations A is wrong 4 times, B can only be right in 3/4 of the remainder, or 9 times in 16. Again, if we have the Approximate Generalisations, 'Most attempts to reduce wages are met by strikes,' and 'Most strikes are successful,' and learn, on statistical inquiry, that in every hundred attempts to reduce wages there are 80 strikes, and that 70 p.c. of the strikes are successful, then 56 p.c. of attempts to reduce wages are unsuccessful.

Of course this method of calculation cannot be quantitatively applied if no statistics are obtainable, as in the testimony of witnesses; and even if an average numerical value could be attached to the evidence of a certain class of witnesses, it would be absurd to apply it to the evidence of any particular member of the class without taking account of his education, interest in the case, prejudice, or general capacity. Still, the numerical illustration of the rapid deterioration of hearsay evidence, when less than quite veracious, puts us on our guard against rumour. To retail rumour may be as bad as to invent an original lie.

(4) If an event may coincide with two or more other independent events, the probability that they will together be a sign of it, is found by multiplying together the fractions representing the improbability that each is a sign of it, and subtracting the product from unity.

This is the rule for estimating the cogency of circumstantial evidence and analogical evidence; or, generally, for combining approximate generalisations "self-corroboratively." If, for example, each of two independent circumstances, A and B, indicates a probability of 6 to 1 in favour of a certain event; taking 1 to represent certainty, 1-6/7 is the improbability of the event, notwithstanding each circumstance. Then $1/7 \times 1/7 = 1/49$, the improbability of both proving it. Therefore the probability of the event is 48 to 1. The matter may be plainer if put thus: A's indication is right 6 times in 7, or 42 in 49; in the remaining 7 times in 49, B's indication will be right 6 times. Therefore, together they will be right 48 times in 49. If each of two witnesses is truthful 6 times in 7, one or the other will be truthful 48 times in 49. But they will not be believed unless they agree; and in the 42 cases of A being right, B will contradict him 6 times; so that they only concur in being right 36 times. In the remaining 7 times in which A is wrong, B will contradict him 6 times, and once they will both be wrong. It does not follow that when both are wrong they will concur; for they may tell very different stories and still contradict one another.

If in an analogical argument there were 8 points of comparison, 5 for and 3 against a certain inference, and the probability raised by each point could be quantified, the total value of the evidence might be estimated by doing similar sums for and against, and subtracting the unfavourable from the favourable total.

When approximate generalisations that have not been precisely quantified combine their evidence, the cogency of the argument increases in the same way, though it cannot be made so definite. If it be true that most poets are irritable, and also that most invalids are irritable, a still greater proportion will be irritable of those who are both invalids and poets.

On the whole, from the discussion of probabilities there emerge four principal cautions as to their use: Not to make a pedantic parade of numerical probability, where the numbers have not been ascertained; Not to trust to our feeling of what is likely, if statistics can be obtained; Not to apply an average probability to special classes or individuals without inquiring whether they correspond to the average type; and Not to trust to the empirical probability of events, if their causes can be discovered and made the basis of reasoning which the empirical probability may be used to verify.

The reader who wishes to pursue this subject further should read a work to which the foregoing chapter is greatly indebted, Dr. Venn's *Logic of Chance*.

CHAPTER XXI
DIVISION AND CLASSIFICATION

§ 1. Classification, in its widest sense, is a mental grouping of facts or phenomena according to their resemblances and differences, so as best to serve some purpose. A "mental grouping": for although in museums we often see the things themselves arranged in classes, yet such an arrangement only contains specimens representing a classification. The classification itself may extend to innumerable objects most of which have never been seen at all. Extinct animals, for example, are classified from what we know of their fossils; and some of the fossils may be seen arranged in a museum; but the animals themselves have disappeared for many ages.

Again, things are classed according to their resemblances and differences: that is to say, those that most closely resemble one another are classed together on that ground; and those that differ from one another in important ways, are distributed into other classes. The more the things differ, the wider apart are their classes both in thought and in the arrangements of a museum. If their differences are very great, as with animals, vegetables and minerals, the classing of them falls to different departments of thought or science, and is often represented in different museums, zoological, botanical, mineralogical.

We must not, however, suppose that there is only one way of classifying things. The same objects may be classed in various ways according to the purpose in view. For gardening, we are usually content to classify plants into trees, shrubs, flowers, grasses and weeds; the ordinary crops of English agriculture are distinguished, in settling their rotation, into white and green; the botanist divides the higher plants into gymnosperms and angiosperms, and the latter into monocotyledons and dicotyledons. The principle of resemblance and difference is recognised in all these cases; but what resemblances or differences are important depends upon the purpose to be served.

Purposes are either (α) special or practical, as in gardening or hunting, or (β) general or scientific, as in Botany or Zoology. The scientific purpose is merely knowledge; it may indeed subserve all particular or practical ends, but has no other end than knowledge directly in view. And whilst, even for knowledge, different classifications may be suitable for different lines of inquiry, in Botany and Zoology the Morphological Classification is that which gives the most general and comprehensive knowledge (see Huxley, *On the Classification of Animals*, ch. 1). Most of what a logician says about classification is applicable to the practical kind; but the scientific (often called 'Natural Classification'), as the most thorough and comprehensive, is what he keeps most constantly before him.

Scientific classification comes late in human history, and at first works over earlier classifications which have been made by the growth of intelligence, of language, and of the practical arts. Even in the distinctions recognised by animals, may be traced the grounds of classification: a cat does not confound a dog with one of its own species, nor water with milk, nor cabbage with fish. But it is in the development of language that the progress of instinctive classification may best be seen. The use of general names implies the recognition of classes of things corresponding to them, which form their denotation, and whose resembling qualities, so far as recognised, form their connotation; and such names are of many degrees of generality. The use of abstract names shows that the objects classed have also been analysed, and that their resembling qualities have been recognised amidst diverse groups of qualities.

Of the classes marked by popular language it is worth while to distinguish two sorts (*cf.* chap. xix. § 4): Kinds, and those having but few points of agreement.

But the popular classifications, made by language and the primitive arts, are very imperfect. They omit innumerable things which have not been found useful or noxious, or have been inconspicuous, or have not happened to occur in the region inhabited by those who speak a particular language; and even things recognised and named may have been very superficially examined, and therefore wrongly classed, as when a whale or porpoise is called a fish, or a slowworm is confounded with snakes. A scientific classification, on the other hand, aims at the utmost comprehensiveness, ransacking the whole world from the depths of the earth to the remotest star for new objects, and scrutinising everything with the aid of crucible and dissecting knife, microscope and spectroscope, to find the qualities and constitution of everything, in order that it may be classed among those things with which it has most in common and distinguished from those other things from which it differs. A scientific classification continually grows more comprehensive, more discriminative, more definitely and systematically coherent. Hence the uses of classification may be easily perceived.

§ 2. The first use of classification is the better understanding of the facts of Nature (or of any sphere of practice); for understanding consists in perceiving and comprehending the likeness and difference of things, in assimilating and distinguishing them; and, in carrying out this process systematically, new correlations of properties are continually disclosed. Thus classification is closely analogous to explanation. Explanation has been shown (chap. xix. § 5) to consist in the discovery of the laws or causes of changes in Nature; and laws and causes imply similarity, or like changes under like conditions: in the same way classification consists in the discovery of resemblances in the things that undergo change. We may say (subject to subsequent qualifications) that Explanation deals with Nature in its dynamic, Classification in its static aspect. In both cases we have a feeling of relief. When the cause of any event is pointed out, or an object is assigned its place in a system of classes, the gaping wonder, or confusion, or perplexity, occasioned by an unintelligible thing, or by a multitude of such things, is dissipated. Some people are more than others susceptible of this pleasure and fastidious about its purity.

A second use of classification is to aid the memory. It strengthens memory, because one of the conditions of our recollecting things is, that they resemble what we last thought of; so that to be accustomed to study and think of things in classes must greatly facilitate recollection. But, besides this, a classification enables us easily to run over all the contrasted and related things that we want to think of. Explanation and classification both tend to rationalise the memory, and to organise the mind in correspondence with Nature.

Every one knows how a poor mind is always repeating itself, going by rote through the same train of words, ideas, actions; and that such a mind is neither interesting nor practical. It is not practical, because the circumstances of life are rarely exactly repeated, so that for a present purpose it is rarely enough to remember only one former case; we need several, that by comparing (perhaps automatically) their resemblances and differences with the one before us, we may select a course of action, or a principle, or a parallel, suited to our immediate needs. Greater fertility and flexibility of thought seem naturally to result from the practice of explanation and classification. But it must be honestly added, that the result depends upon the spirit in which such study is carried on; for if we are too fond of finality, too eager to believe that we have already attained a greater precision and comprehension than are in fact attainable, nothing can be more petrific than 'science,' and our last state may be worse than the first. Of this, students of Logic have often furnished examples.

§ 3. Classification may be either Deductive or Inductive; that is to say, in the formation of classes, as in the proof of propositions, we may, on the whole, proceed from the more to the less, or from the less to the more general; not that these two processes are entirely independent.

If we begin with some large class, such as 'Animal,' and subdivide it deductively into Vertebrate and Invertebrate, yet the principle of division (namely, central structure) has first been reached by a comparison of examples and by generalisation; if, on the other hand, beginning with individuals, we group them inductively into classes, and these again into wider ones (as dogs, rats, horses, whales and monkeys into mammalia) we are guided both in special cases by hypotheses as to the best grounds of resemblance, and throughout by the general principle of classification—to associate things that are alike and to separate things that are unlike. This principle holds implicitly a place in classification similar to that of causation in explanation; both are principles of intelligence. Here, then, as in proof, induction is implied in deduction, and deduction in induction. Still, the two modes of procedure may be usefully distinguished: in deduction, we proceed from the idea of a whole to its parts, from general to special; in induction, from special (or particular) to general, from parts to the idea of a whole.

§ 4. The process of Deductive Classification, or Formal Division, may be represented thus:

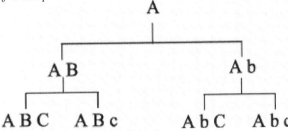

Given any class (A) to be divided:

1. Select one important character, attribute, or quality (B), not common to all the individuals comprehended in the class, as the basis of division (*fundamentum divisionis*).

2. Proceed by Dichotomy; that is, cut the given class into two, one having the selected attribute (say, B), the other not having it (b). This, like all formal processes, assumes the principles of Contradiction and Excluded Middle, that 'No A is both B and not-B,' and that 'Every A is either B or not-B' (chap. vi. § 3); and if these principles are not true, or not applicable, the method fails.

When a class is thus subdivided, it may be called, in relation to its subclasses, a Genus; and in relation to it, the subclasses may be called Species: thus—genus A, species AB and Ab, *etc.*

3. Proceed gradually in the order of the importance of characters; that is, having divided the given class, subdivide on the same principle the two classes thence arising; and so again and again, step by step, until all the characters are exhausted: *Divisio ne fiat per saltum*.

Suppose we were to attempt an exhaustive classification of things by this method, we must begin with 'All Things,' and divide them (say) into phenomenal and not-phenomenal, and then subdivide phenomena, and so on, thus:

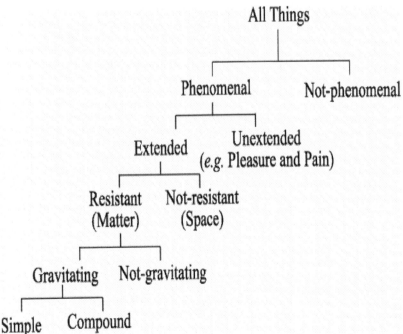

Having subdivided 'Simple' by all possible characters, we must then go back and similarly subdivide Not-phenomenal, Unextended, Not-resistant, Not-gravitating, and Compound. Now, if we knew all possible characters, and the order of their importance, we might prepare *a priori* a classification of all possible things; at least, of all things that come under the principles of Contradiction and Excluded Middle. Many of our compartments might contain nothing actual; there may, for example, be nothing that is not phenomenal to some mind, or nothing that is extended and not-resistant (no vacuum), and so forth. This would imply a breach of the rule, that the dividing quality be not common to the whole class; but, in fact, doubts have been, and are, seriously entertained whether these compartments are filled or not. If they are not, we have concepts representing nothing, which have been generated by the mere force of grammatical negation, or by the habit of thinking according to the principle of Excluded Middle; and, on the strength of these empty concepts, we have been misled into dividing by an attribute, which (being universal) cannot be a *fundamentum divisionis*. But though in such a classification places might be empty, there would be a place for everything; for whatever did not come into some positive class (such as Gravitating) must fall under one of the negative classes (the 'Nots') that run down the right-hand side of the Table and of its subdivisions.

This is the ideal of classification. Unfortunately we have to learn what characters or attributes are possible, by experience and comparison; we are far from knowing them all: and we do not know the order of their importance; nor are we even clear what 'important' means in this context, whether 'widely prevalent,' or 'ancient,' or 'causally influential,' or 'indicative of others.' Hence, in classifying actual things, we must follow the inductive method of beginning with particulars, and sorting them according to their likeness and difference as discovered by investigation. The exceptional cases, in which deduction is really useful, occur where certain limits to the number and combination of qualities happen to be known, as they may be in human institutions, or where there are mathematical conditions. Thus, we might be able to classify orders of Architecture, or the classical metres and stanzas of English poetry; though, in fact, these things are too free, subtle and complex for deductive treatment: for do not the Arts grow like trees? The only sure cases are mathematical; as we may show that there are possible only three kinds of plane triangles, four conic sections, five regular solids.

§ 5. The rules for *testing* a Division are as follows:

1. Each Sub-class, or Species, should comprise less than the Class, or Genus, to be divided. This provides that the division shall be a real one, and not based upon an attribute common to the whole class; that, therefore, the first rule for making a division shall have been adhered to. But, as in § 4, we are here met by a logical difficulty. Suppose that the class to be divided is A, and that we attempt to divide upon the attribute B, into AB and Ab; is this a true division, if we do not know any A that is not B? As far as our knowledge extends, we have not divided A at all. On the other hand, our knowledge of concrete things is never exhaustive; so that, although we know of no A that is not B, it may yet exist, and we have seen that it is a logical caution not to assume what we do not know. In a deductive classification, at least, it seems better to regard every attribute as a possible ground of division. Hence, in the above division of 'All Things,'—'Not-phenomenal,' 'Extended-Not-resistant,' 'Resistant-Not-gravitating,' appear as negative classes (that is, classes based on the negation of an attribute), although their real existence may be doubtful. But, if this be justifiable, we must either rewrite the first test of a division thus: 'Each sub-class should *possibly* comprise less than the class to be divided'; or else we must confine the test to (*a*) thoroughly empirical divisions, as in dividing Colour into Red and Not-red, where we know that both sub-classes are real; and (*b*) divisions under demonstrable conditions—as in dividing the three kinds of triangles by the quality equilateral, we know that it is only applicable to acute-angled triangles, and do not attempt to divide the right-angled or obtuse-angled by it.

2. The Sub-classes taken together should be equal to the Class to be divided: the sum of the Species constitutes the Genus. This provides that the division shall be exhaustive; which dichotomy always secures, according to the principle of Excluded Middle; because whatever is not in the positive class, must be in the negative: Red and Not-red include all colours.

3. The Sub-classes must be opposed or mutually exclusive: Species must not overlap. This again is secured by dichotomy, according to the principle of Contradiction, provided the division be made upon one attribute at a time. But, if we attempt to divide simultaneously upon two attributes, as 'Musicians' upon 'nationality' and 'method,' we get what is called a Cross-division, thus 'German Musicians.' 'Not-German,' 'Classical,' 'Not-Classical;' for these classes may overlap, the same men sometimes appearing in two groups—Bach in 'German' and 'Classical,' Pergolesi in 'Not-German' and 'Classical.' If, however, we divide Musicians upon these attributes successively, cross division will be avoided, thus:

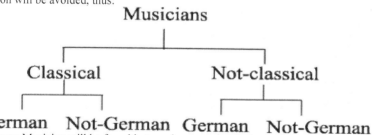

Here no Musician will be found in two classes, *unless* he has written works in two styles, *or unless* there are works whose style is undecided. This "unless—or unless" may suggest caution in using dichotomy as a short cut to the classification of realities.

4. No Sub-class must include anything that is not comprised in the class to be divided: the Genus comprises all the Species. We must not divide Dogs into fox-terriers and dog-fish.

§ 6. The process of Inductive Classification may be represented thus:

Given any multitude of individuals to be classified:

(1) Place together in groups (or in thought) those things that have in common the most, the most widely diffused and the most important qualities.

(2) Connect those groups which have, as groups, the greater resemblance, and separate those that have the greater difference.

(3) Demarcate, as forming higher or more general classes, those groups of groups that have important characters in common; and, if possible, on the same principle, form these higher classes into classes higher still: that is to say, graduate the classification upwards.

Whilst in Division the terms 'Genus' and 'Species' are entirely relative to one another and have no fixed positions in a gradation of classes, it has been usual, in Inductive Classification, to confine the term 'Species' to classes regarded as lowest in the scale, to give the term 'Genera' to classes on the step above, and at each higher step to find some new term such as 'Tribe,' 'Order,' 'Sub-kingdom,' 'Kingdom';

as may be seen by turning to any book on Botany or Zoology. If, having fixed our Species, we find them subdivisible, it is usual to call the Sub-species 'Varieties.'

Suppose an attempt to classify by this method the objects in a sitting-room. We see at a glance carpets, mats, curtains, grates, fire-irons, coal-scuttles, chairs, sofas, tables, books, pictures, musical instruments, *etc.* These may be called 'Species.' Carpets and mats go together; so do chairs and sofas; so do grates, fire-irons, and coal-scuttles and so on. These greater groups, or higher classes, are 'Genera.' Putting together carpets, mats and curtains as 'warmth-fabrics'; chairs, sofas and tables as 'supports'; books, pictures and musical instruments as 'means of culture'; these groups we may call Orders. Sum up the whole as, from the housewife's point of view, 'furniture.' If we then subdivide some of the species, as books into poetry, novels, travels, *etc.*, these Sub-species may be considered 'Varieties.'

A Classification thus made, may be tested by the same rules as those given for testing a Division; but if it does not stand the test, we must not infer that the classification is a bad one. If the best possible, it is good, though formally imperfect: whatever faults are found must then be charged upon the 'matter,' which is traditionally perverse and intractable. If, for example, there is a hammock in the room, it must be classed not with the curtains as a warmth-fabric, but with the sofas as a support; and books and pictures may be classed as, in a peculiar sense, means of culture, though all the objects in the room may have been modified and assorted with a view to gratifying and developing good taste.

§ 7. The difficulty of classifying natural objects is very great. It is not enough to consider their external appearance: exhaustive knowledge of their internal structure is necessary, and of the functions of every part of their structure. This is a matter of immense research, and has occupied many of the greatest minds for very many years. The following is a tabular outline of the classification of the

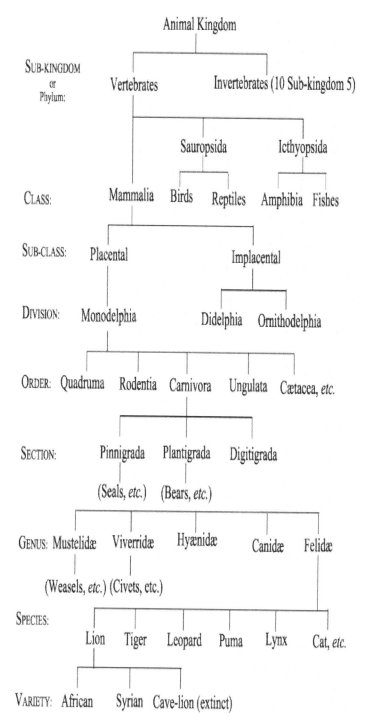

As there is not space enough to tabulate such a classification in full, I have developed at each step the most interesting groups: Vertebrates, Mammalia, Monodelphia Carnivora, Digitigrada, Felidæ, Lion. Most of the other groups in each grade are also subdivisible, though some of them contain far fewer sub-classes than others.

To see the true character of this classification, we must consider that it is based chiefly upon knowledge of existing animals. Some extinct animals, known by their fossils, find places in it; for others new places have been made. But it represents, on the whole, a cross-section, or cross-sections of Nature as developing in time; and, in order to give a just view of the relations of animals, it must be seen in the light of other considerations. The older systems of classification, and the rules for making them, seem to have assumed that an actual system of classes, or of what Mill calls 'Kinds,' exists in nature, and that the relations of Kinds in this system are determined by quantity of resemblance in co-inherent qualities, as the ground of their affinity.

§ 8. Darwin's doctrine of the origin of species affects the conception of natural classification in several ways, (1) If all living things are blood-relations, modified in the course of ages according to their various conditions of life, 'affinity' must mean 'nearness of common descent'; and it seems irrational to propose a classification upon any other basis. We have to consider the Animal (or the Vegetable)

Kingdom as a family tree, exhibiting a long line of ancestors, and (descended from them) all sorts of cousins, first, second, third, *etc.*, perhaps once, twice, or oftener 'removed.' Animals in the relation of first cousins must be classed as nearer than second cousins, and so on.

But, if we accept this principle, and are able to trace relationship, it may not lead to the same results as would be reached by simply relying upon the present 'quantity of resemblance,' unless we understand this in a very particular way. For the most obvious features of an animal may have been recently acquired; which often happens with those characters that adapt an animal to its habits of life, as the wings of a bat, or the fish-like shape of a dolphin; or as in cases of 'mimicry.' Some butterflies, snakes, *etc.*, have grown to resemble closely, in a superficial way, other butterflies and snakes, from which a stricter investigation widely separates them; and this superficial resemblance is probably a recent acquisition, for the sake of protection; the imitated butterflies being nauseous, and the imitated snakes poisonous. On the other hand, ancient and important traits of structure may, in some species, have dwindled into inconspicuous survivals or be still found only in the embryo; so that only great knowledge and sagacity can identify them; yet upon ancient traits, though hidden, classification depends. The seal seems nearer allied to the porpoise than to the tiger, the shrew nearer to the mouse than to the hedgehog; and the Tasmanian wolf looks more like a true wolf, the Tasmanian devil more like a badger, than like a kangaroo: yet the seal is nearer akin to the tiger, the shrew to the hedgehog, and the Tasmanian flesh-eaters are marsupial, like the kangaroo. To overcome this difficulty we must understand the resemblance upon which classification is based to include resemblance of Causation, that is, the fact itself of descent from common ancestors. For organic beings, all other rules of classification are subordinate to one: trace the genealogy of every form.

By this rule we get a definite meaning for the phrase 'important or fundamental attribute' as determining organic classes; namely, most ancient, or 'best serving to indicate community of origin.' Grades of classification will be determined by such fundamental characters, and may correspond approximately to the more general types (now extinct) from which existing animals have descended.

(2) By the hypothesis of development the fixity of species is discredited. The lowest grade of a classification is made up not of well-defined types unchanging from age to age, but of temporary species, often connected by uncertain and indistinct varieties: some of which may, in turn, if the conditions of their existence alter, undergo such changes as to produce new species. Hence the notion that Kinds exist in organic nature must be greatly modified. During a given period of a few thousand years, Kinds may be recognised, because, under such conditions as now prevail in the world, that period of time is insufficient to bring about great changes. But, if it be true that lions, tigers, and leopards have had a common ancestor, from whose type they have gradually diverged, it is plain that their present distinctness results only from the death of intermediate specimens and the destruction of intermediate varieties. Were it possible to restore, by the evidence of fossils, all the ranks of the great processions that have descended from the common ancestor, there would nowhere occur a greater difference than between offspring and parents; and the appearance of Kinds existing in nature, which is so striking in a museum or zoological garden, would entirely vanish.

A classification, then, as formerly observed, represents a cross-section of nature as developing in time: could we begin at the beginning and follow this development down the course of time, we should find no classes, but an ever-moving, changing, spreading, branching continuum. It may be represented thus: Suppose an animal (or plant) A, extending over a certain geographical area, subject to different influences and conditions of climate, food, hill and plain, wood and prairie, enemies and rivals, and undergoing modifications here and there in adaptation to the varying conditions of life: then varieties appear. These varieties, diverging more and more, become distinct species (AB, AC, AD, AX). Some of these species, the more widely diffused, again produce varieties; which, in turn become species (ABE, ABF, ADG, ADH). From these, again, ABE, ABFI, ABFJ, AC, ADHK, ADHL, ADHM, the extant species, descend.

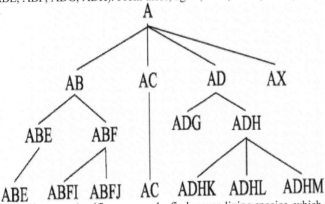

If in this age a classifier appears, he finds seven living species, which can be grouped into four genera (ABE, ABF, AC, ADH), and these again into three Families (AB, AC, AD), all forming one Order. But the animals which were their ancestors are all extinct. If the fossils of any of them—say AB, ADG and AX—can be found, he has three more species, one more genus (ADG), and one more family (AX). For AC, which has persisted unchanged, and AX, which has become extinct, are both of them Families, each represented by only one species. It seems necessary to treat such ancient types as species on a level with extant forms; but the naturalist draws our attention to their archaic characteristics, and tries to explain their places in the order of evolution and their relationships.

But now suppose that he could find a fossil specimen of every generation (hundreds of thousands of generations), from ABFI, *etc.*, up to A; then, as each generation would only differ from the preceding as offspring from parents, he would be unable at any point to distinguish a species; at most, he would observe a slightly marked variety. ABFI and ABFJ would grow more and more alike, until they became indistinguishable in ABF; ABF and ABE would merge into AB; AB, AC, AD and AX would merge into A. Hence, the appearance of species is due to our taking cross-sections of time, or comparing forms that belong to periods remote from one another (like AX, ADG, and ADHK, or AD, ADH and ADHK), and this appearance of species depends upon the destruction of ancestral intermediate forms.

(3) The hypothesis of development modifies the logical character of classification: it no longer consists in a direct induction of co-inherent characters, but is largely a deduction of these from the characters of earlier forms, together with the conditions of variation; in

other words, the definition of a species must, with the progress of science, cease to be a mere empirical law of co-inherence and become a derivative law of Causation. But this was already implied in the position that causation is the fundamental principle of the explanation of concrete things; and accordingly, the derivative character of species or kinds extends beyond organic nature.

§ 9. The classification of inorganic bodies also depends on causation. There is the physical classification into Solids, Liquids, and Gases. But these states of matter are dependent on temperature; at different temperatures, the same body may exist in all three states. They cannot therefore be defined as solid, liquid, or gaseous absolutely, but only within certain degrees of temperature, and therefore as dependent upon causation. Similarly, the geological classification of rocks, according to relative antiquity (primary, secondary, tertiary, with their subdivisions), and mode of formation (igneous and aqueous), rests upon causation; and so does the chemical classification of compound bodies according to the elements that enter into them in definite proportions. Hence, only the classification of the elements themselves (amongst concrete things), at present, depends largely upon empirical Coinherence. If the elements remain irresolvable into anything simpler, the definitions of the co-inherent characters that distinguish them must be reckoned amongst the ultimate Uniformities of Nature. But if a definite theory of their origin both generally and severally, whether out of ether-vortices, or groups of electric corpuscles, or whatnot, shall ever gain acceptance, similarity of genesis or causation will naturally be the leading consideration in classifying the chemical elements. To find common principles of causation, therefore, constitutes the verification of every Natural Classification. The ultimate explanation of nature is always causation; the Law of Causation is the backbone of the system of Experience.

CHAPTER XXII

NOMENCLATURE, DEFINITION, PREDICABLES

§ 1. Precision of thought needs precision of language for the recording of such thought and for communicating it to others. We can often remember with great vividness persons, things, landscapes, changes and actions of persons or things, without the aid of language (though words are often mixed with such trains of imagery), and by this means may form judgments and inferences in particular cases; but for general notions, judgments and inferences, not merely about this or that man, or thing, but about all men or all kinds of things, we need something besides the few images we can form of them from observation. Even if we possess generic images, say, of 'horse' or 'cat' (that is, images formed, like composite photographs, by a coalescence of the images of all the horses or cats we have seen, so that their common properties stand out and their differences frustrate and cancel one another), these are useless for precise thought; for the generic image will not correspond with the general appearance of horse or cat, unless we have had proportional experience of all varieties and have been impartially interested in all; and, besides, what we want for general thought is not a generic image of the appearance of things, though it were much more definite and fairly representative than such images ever are, but a general representation of their important characters; which may be connected with internal organs, such as none but an anatomist ever sees. We require a symbol connected with the general character of a thing, or quality, or process, as scientifically determined, whose representative truth may be trusted in ordinary cases, or may be verified whenever doubt arises. Such symbols are for most purposes provided by language; Mathematics and Chemistry have their own symbols.

§ 2. First there should be "a name for every important meaning": (*a*) A Nomenclature, or system of the names of all classes of objects, adapted to the use of each science. Thus, in Geology there are names for classes of rocks and strata, in Chemistry for the elements and their compounds, in Zoology and Botany for the varieties and species of animals and plants, their genera, families and orders.

To have such names, however, is not the whole aim in forming a scientific language; it is desirable that they should be systematically significant, and even elegant. Names, like other instruments, ought to be efficient, and the efficiency of names consists in conveying the most meaning with the least effort. In Botany and Zoology this result is obtained by giving to each species a composite name which includes that of the genus to which it belongs. The species of Felidæ given in chap. xvii. § 7, are called *Felis leo* (lion), *Felis tigris* (tiger), *Felis leopardus* (leopard), *Felis concolor* (puma), *Felis lyncus* (European lynx), *Felis catus* (wild cat). In Chemistry, the nomenclature is extremely efficient. Names of the simpler compounds are formed by combining the names of the elements that enter into them; as Hydrogen Chloride, Hydrogen Sulphide, Carbon Dioxide; and these can be given still more briefly and efficiently in symbols, as HCl, H_2S, CO_2. The symbolic letters are usually initials of the names of the elements: as C = Carbon, S = Sulphur; sometimes of the Latin name, when the common name is English, as Fe = Iron. Each letter represents a fixed quantity of the element for which it stands, *viz.*, the atomic weight. The number written below a symbol on the right-hand side shows how many atoms of the element denoted enter into a molecule of the compound.

(*b*) A Terminology is next required, in order to describe and define the things that constitute the classes designated by the nomenclature, and to describe and explain their actions.

(i) A name for every integral part of an object, as head, limb, vertebra, heart, nerve, tendon; stalk, leaf, corolla, stamen, pistil; plinth, frieze, *etc.* (ii) A name for every metaphysical part or abstract quality of an object, and for its degrees and modes; as extension, figure, solidity, weight; rough, smooth, elastic, friable; the various colours, red, blue, yellow, in all their shades and combinations and so with sounds, smells, tastes, temperatures. The terms of Geometry are employed to describe the modes of figure, as angular, curved, square, elliptical; and the terms of Arithmetic to express the degrees of weight, elasticity, temperature, pitch of sound. When other means fail, qualities are suggested by the names of things which exhibit them in a salient way; figures by such terms as amphitheatre, bowl-like, pear-shaped, egg-shaped; colours by lias-blue, sky-blue, gentian-blue, peacock-blue; and similarly with sounds, smells and tastes. It is also important to express by short terms complex qualities, as harmony, fragrance, organisation, sex, symmetry, stratification.

(iii) In the explanation of Nature we further require suitable names for processes and activities: as deduction, conversion, verification, addition, integration, causation, tendency, momentum, gravitation, aberration, refraction, conduction, affinity, combination, germination, respiration, attention, association, development.

There may sometimes be a difficulty in distinguishing the terms which stand for qualities from those that express activities, since all qualities imply activities: weight, for example, implies gravitation; and the quality heat is also a kind of motion. The distinction aimed at lies between a quality as perceived by means of an effect upon our senses (as weight is resistance to our effort in lifting; heat, a sensation when we approach fire), and that property of a body which is conceived to account for its energy (as gravitation that brings a body to the ground, or physical heat that expands an iron bar or works an engine). The former class of words, expressing qualities, are chiefly used in description: the latter class, expressing activities, are chiefly needed in explanation. They correspond respectively, like classification and explanation, with the static and dynamic aspects of Nature.

The terms of ordinary language fall into the same classes as those of science: they stand for things, classes of things, parts, or qualities, or activities of things; but they are far less precise in their signification. As long as popular thought is vague its language must be vague; nor is it desirable too strictly to correct the language whilst the thought is incorrigible. Much of the effect of poetry and eloquence depends upon the elasticity and indirect suggestiveness of common terms. Even in reasoning upon some subjects, it is a mistake to aim at an unattainable precision. It is better to be vaguely right than exactly wrong. In the criticism of manners, of fine art, or of literature, in politics, religion and moral philosophy, what we are anxious to say is often far from clear to ourselves; and it is better to indicate our meaning approximately, or as we feel about it, than to convey a false meaning, or to lose the warmth and colour that are the life of such reflections. It is hard to decide whether more harm has been done by sophists who take a base advantage of the vagueness of common terms, or by honest paralogists (if I may use the word) who begin by deceiving themselves with a plausible definiteness of expression, and go on to propagate their delusions amongst followers eager for systematic insight but ignorant of the limits of its possibility.

§ 3. A Definition is necessary (if possible) for every scientific name. To define a name is to give a precise statement of its meaning or connotation. The name to be defined is the subject of a proposition, whose predicate is a list of the fundamental qualities common to the things or processes which the subject denotes, and on account of possessing which qualities this name is given to them.

Thus, a curve is a line of which no part is straight. The momentum of a moving body is the product of its mass and its velocity (these being expressed in numbers of certain units). Nitrogen is a transparent colourless gas, atomic weight 14, specific gravity .9713, not readily combining, *etc*. A lion is a monodelphian mammal, predatory, walking on its toes, of nocturnal habits, with a short rounded head and muzzle; dental formula: Incisors (3-3)/(3-3), canines (1-1)/(1/1), præmolars (3-3)/(2-2), molars (1-1)/(1-1) = 30; four toes on the hind and five on the fore foot, retractile claws, prickly tongue, light and muscular in build, about 9½ feet from muzzle to tip of tail, tawny in colour, the males maned, with a tufted tail. If anything answers to this description, it is called a lion; if not, not: for this is the meaning of the name.

For ordinary purposes, it may suffice to give an Incomplete Definition; that is, a list of qualities not exhaustive, but containing enough to identify the things denoted by the given name; as if we say that a lion is 'a large tawny beast of prey with a tufted tail.' Such purposes may also be served by a Description; which is technically, a proposition mentioning properties sufficient to distinguish the things denoted, but not the properties that enter into the definition; as if nitrogen be indicated as the gas that constitutes 4/5 of the atmosphere.

§ 4. The rules for testing a Definition are: I.—As to its Contents—
(1) It must state the whole connotation of the name to be defined.
(2) It must not include any quality derivative from the connotation. Such a quality is called a Proprium. A breach of this rule can do no positive harm, but it is a departure from scientific economy. There is no need to state in the definition what can be derived from it; and whatever can be derived by causation, or by mathematical demonstration, should be exhibited in that manner.
(3) It must not mention any circumstance that is not a part of the connotation, even though it be universally found in the things denoted. Such a circumstance, if not derivable from the connotation, is called an Accident. That, for example, the lion at present only inhabits the Old World, is an accident: if a species otherwise like a lion were found in Brazil, it would not be refused the name of lion on the score of locality. Whilst, however, the rules of Logic have forbidden the inclusion of proprium or accident in a definition, in fact the definitions of Natural History often mention such attributes when characteristic. Indeed, definitions of superordinate classes—Families and Orders—not infrequently give qualities as generally found in the subordinate classes, and at the same time mention exceptional cases in which they do not occur.

II.—As to its Expression—
(4) A Definition must not include the very term to be defined, nor any cognate. In defining 'lion' we must not repeat 'lion,' nor use 'leonine'; it would elucidate nothing.
(5) It must not be put in vague language.
(6) It must not be in a negative form, if a positive form be obtainable. We must not be content to say that a lion is 'no vegetarian,' or 'no lover of daylight.' To define a curve as a line 'always changing its direction' may be better than as 'in no part straight.'

§ 5. The process of determining a Definition is inseparable from classification. We saw that classification consists in distributing things into groups according to their likenesses and differences, regarding as a class those individuals which have most qualities in common. In doing so we must, of course, recognise the common qualities or points of likeness; and to enumerate these is to define the name of the class. If we discover the qualities upon which a class is based by direct observation and induction, by the same method we discover the definition of its name.

We saw also that classification is not merely the determination of isolated groups of things, but a systematic arrangement of such groups in relation to one another. Hence, again, Definitions are not independent, but relative to one another; and, of course, in the same way as classes are relative. That is to say, as a class is placed in subordination to higher or more comprehensive groups, so the definition of its name is subordinate to that of their names; and as a class stands in contrast with co-ordinate classes (those that are in the same degree of subordination to the same higher groups), so the definition of its name is in contrast or co-ordination with the definitions of their names. Lion is subordinate to *Felis*, to Digitigrade, to Carnivore and so on up to Animal; and, beyond the Animal Kingdom, to Phenomenon; it is co-ordinate with tiger, puma, *etc.*; and more remotely it is co-ordinate with dog, jackal, wolf, which come under *Canis*—a genus co-ordinate with *Felis*. The definition of lion, therefore, is subordinate to that of *Felis*, and to all above it up to Phenomenon; and is co-ordinate with that of tiger, and with all species in the same grade. This is the ground of the old method of definition *per genus et differentiam*.

The genus being the next class above any species, the *differentia* or Difference consists of the qualities which mark that species in addition to those that mark the genus, and which therefore distinguish it from all other species of the same genus. In the above definition of lion, for example, all the properties down to "light and muscular in build" are generic, that is, are possessed by the whole genus, *Felis*; and the remaining four (size, colour, tufted tail, and mane in the male) are the Difference or specific properties, because in those points the lion contrasts with the other species of that genus. Differences may be exhibited thus:

	Lion.	Tiger.
Size:	about 9½ feet from nose to tip of tail.	About 10 feet.
Colour:	tawny.	Warm tawny, striped with black.
Tail:	tufted.	Tapering.
Mane:	present in the male.	Both sexes maneless.

There are other differences in the shape of the skull. In defining lion, then, it would have been enough to mention the genus and the properties making up the Difference; because the properties of the genus may be found by turning to the definition of the genus; and, on the principle of economy, whatever it is enough to do it is right to do. To define 'by genus and difference' is a point of elegance, when the genus is known; but the only way of knowing it is to compare the individuals comprised in it and in co-ordinate genera, according to the methods of scientific classification. It may be added that, as the genus represents ancestral derivation, the predication of genus in a definition indicates the remote causes of the phenomena denoted by the name defined. And this way of defining corresponds with the method of double naming by genus and species: *Felis leo, Felis tigris, etc.*; *Vanessa Atalanta, Vanessa Io, etc.*

The so-called Genetic Definition, chiefly used in Mathematics, is a rule for constructing that which a name denotes, in such a way as to ensure its possessing the tributes connoted by the name. Thus, for a circle: Take any point and, at any constant distance from it, trace a line returning into itself. In Chemistry a genetic definition of any compound might be given in the form of directions for the requisite synthesis of elements.

§ 6. The chief difficulty in the definition of scientific names consists in determining exactly the nature of the things denoted by them, as in classifying plants and animals. If organic species are free growths, continually changing, however gradually, according as circumstances give some advantage to one form over others, we may expect to find such species branching into varieties, which differ considerably from one another in some respects, though not enough to constitute distinct species. This is the case; and, consequently, there arises some uncertainty in collecting from all the varieties those attributes which are common to the species as a whole; and, therefore, of course, uncertainty in defining the species. The same difficulty may occur in defining a genus, on account of the extent to which some of its species differ from others, whilst having enough of the common character to deter the classifier from forming a distinct genus on their account. On the other hand the occurrence of numerous intermediate varieties may make it difficult to distinguish genera or species at all. Even the Kingdoms of plants and animals are hard to discriminate at the lowest levels of organisation. Now, where there is a difficulty of classification there must be a corresponding difficulty of definition.

It has been proposed in such cases to substitute a Type for a Definition; to select some variety of a species, or species of a genus, as exhibiting its character in an eminent degree, and to regard other groups as belonging to the same species or genus, according as they agree more with this type than with other types representing other species or genera. But the selection of one group as typical implies a recognition of its attributes as prevailing generally (though not universally) throughout the species or genus; and to recognise these attributes and yet refuse to enumerate them in a definition, seems to be no great gain. To enumerate the attributes of the type as an Approximate Definition of the species or genus, true of *most* of the groups constituting the species or genus, answers the same purpose, is more explicit, and can mislead no one who really attends to the exposition. An approximate definition is, indeed, less misleading than the indication of a type; for the latter method seems to imply that the group which is now typical has a greater permanence or reality than its co-ordinate groups; whereas, for aught we know, one of the outside varieties or species may even now be superseding and extinguishing it. But the statement of a definition as approximate, is an honest confession that both the definition and the classification are (like a provisional hypothesis) merely the best account we can give of the matter according to our present knowledge.

§ 7. The limits of Definition are twofold: (*a*) A name whose meaning cannot be analysed cannot be defined. This limitation meets us only in dealing with the names of the metaphysical parts or simple qualities of objects under the second requisite of a Terminology. Resistance and weight, colour and its modes, many names of sounds, tastes, smells, heat and cold—in fact, whatever stands for an unanalysable perception, cannot be made intelligible to any one who has not had experience of the facts denoted; they cannot be defined, but only exemplified. A sort of genetic definition may perhaps be attempted, as if we say that colour is the special sensation of the cones of the retina, or that blue is the sensation produced by a ray of light vibrating about 650,000,000,000,000 times a second; but such expressions can give no notion of our meaning to a blind man, or to any one who has never seen a blue object. Nor can we explain what heat is like, or the smell of tobacco, to those who have never experienced them; nor the sound of C 128 to one who knows nothing of the musical scale.

If we distinguish the property of an object from the sensation it excites in us, we may define any simple property as 'the power of producing the sensation'; the colour of a flower as the power of exciting the sensation of colour in us. Still, this gives no information to the blind nor to the colour-blind. Abstract names may be defined by defining the corresponding concrete: the definition of 'human nature' is the same as of 'man.' But if the corresponding concrete be a simple sensation (as blue), this being indefinable, the abstract (blueness) is also indefinable.

(*b*) The second limit of Definition is the impossibility of exhausting infinity, which would be necessary in order to convey the meaning of the name of any individual thing or person. For, as we saw in chap. iv., if in attempting to define a proper name we stop short of infinity, our list of qualities or properties may possibly be found in two individuals, and then it becomes the definition of a class-name or general name, however small the actual class. Hence we can only give a Description of that which a proper name denotes, enumerating enough of its properties to distinguish it from everything else as far as our knowledge goes.

§ 8. The five Predicables (Species, Genus, Difference, Proprium, Accident) may best be discussed in connection with Classification and Definition; and in giving an account of Classification, most of what has to be said about them has been anticipated. Their name, indeed, connects them with the doctrine of Propositions; for Predicables are terms that may be predicated, classified according to their connotative relation to the subject of a proposition (that is, according to the relation in which their connotation stands to the connotation of the subject): nevertheless, the significance of the relations of such predicates to a subject is derivative from the general doctrine of classification.

For example, in the proposition 'X is Y,' Y must be one of the five sorts of predicables in relation to X; but of what sort, depends upon what X (the subject) is, or means. The subject of the proposition must be either a definition, or a general connotative name, or a singular name.

If X be a definition, Y must be a species; for nothing but a general name can be predicated of a definition: and, strictly speaking, it is only in relation to a definition (as subject) that species can be a predicable; when it is called *Species predicabilis* (1).

If X be a connotative name, it is itself a species (*Species subjicibilis*); and the place of the subject of a proposition is the usual one for species. The predicate, Y, may then be related to the species in three different ways. First, it may be a definition, exactly equivalent to the species;—in fact, nothing else than the species in an explicit form, the analysis of its connotation. Secondly, the predicate may be, or connote, some *part only* of the definition or connotation of the species; and then it is either genus (2), or difference (3). Thirdly, the predicate may connote *no part* of the definition, and then it is either derivable from it, being a proprium (4), or not derivable from it, being an accident (5). These points of doctrine will be expanded and illustrated in subsequent pages.

If X be a singular name, deriving connotation from its constituent terms ([chap. iv. § 2](#)), as 'The present Emperor of China,' it may be treated as a *Species subjicibilis*. Then that he is 'an absolute monarch,' predicates a genus; because that is a genus of 'Emperor,' a part of the singular name that gives it connotation. That he wears a yellow robe is a proprium, derivable from the ceremonial of his court. That he is thirty years of age is an accident.

But if X be a proper name, having no connotation, Y must always be an accident; since there can then be no definition of X, and therefore neither species, genus, difference, nor proprium. Hence, that 'John Doe is a man' is an accidental proposition: 'man' is not here a *Species predicabilis*; for the name might have been given to a dog or a mountain. That is what enables the proposition to convey information: it would be useless if the proper name implied 'humanity.'

'Species' is most frequently used (as in Zoology) for the *class denoted* by a general name; but in Logic it is better to treat it as a general name used connotatively for the attributes possessed in common by the things denoted, and on account of which they are regarded as a class: it is sometimes called the Essence ([§ 9](#)). In this connotative sense, a species is implicitly what the definition is explicitly; and therefore the two are always simply convertible. Thus, 'A plane triangle' (species) is 'a figure enclosed by three straight lines' (definition): clearly we may equally say, 'A figure enclosed by three straight lines is a plane triangle.' It is a simple identity.

A genus is also commonly viewed denotatively, as a class containing smaller classes, its species; but in Logic it is, again, better to treat it connotatively, as a name whose definition is part of the definition of a given species.

A difference is the remainder of the definition of any species after subtracting a given genus. Hence, the genus and difference together make up the species; whence the method of definition *per genus et differentiam* (ante, [§ 5](#)).

Whilst in Botany and Zoology the species is fixed at the lowest step of the classification (varieties not being reckoned as classes), and the genus is also fixed on the step next above it, in Logic these predicables are treated as movable up and down the ladder: any lower class being species in relation to any higher; which higher class, wherever taken, thus becomes a genus. Lion may logically be regarded as a species of digitigrade, or mammal, or animal; and then each of these is a genus as to lion: or, again, digitigrade may be regarded as a species of mammal, or mammal as a species of animal. The highest class, however, is never a species; wherefore it is called a *Summum Genus*: and the lowest class is never a genus; wherefore it is called an *Infima Species*. Between these two any step may be either species or genus, according to the relation in which it is viewed to other classes, and is then called Subaltern. The *summum genus*, again, may be viewed in relation to a *given* universe or *suppositio* (that is, any limited area of existence now the object of attention), or to the *whole* universe. If we take the animal kingdom as our *suppositio*, Animal is the *summum genus*; but if we take the whole universe, 'All things' is the *summum genus*.

"Porphyry's tree" is used to illustrate this doctrine. It begins with a *summum genus*, 'Substance,' and descends by adding differences, step by step, to the *infima species*, 'Man.' It also illustrates Division by Dichotomy.

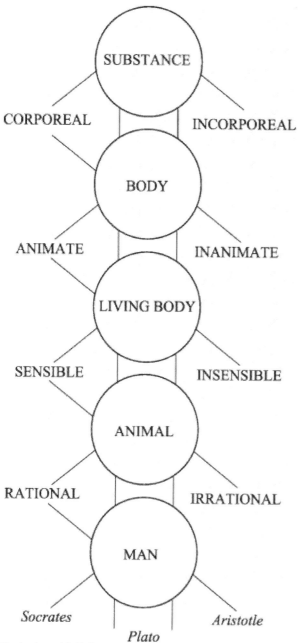

Plato

    Beginning with 'Substance,' as *summum genus*, and adding the difference 'Corporeal,' we frame the species 'Body.' Taking 'Body' as the genus and adding the difference 'Animate,' we frame the species 'Living Body;' and so on till 'Man' is reached; which, being *infima species*, is only subdivisible into individuals. But the division of Man into individuals involves a change of principle; it is a division of the denotation, not an increase of the connotation as in the earlier steps. Only one side of each dichotomy is followed out in the 'tree': if the other side had been taken, Incorporeal Substance would be 'Spirit'; which might be similarly subdivided.

    Genus and species, then, have a double relation. In denotation the genus includes the species; in connotation the species includes the genus. Hence the doctrine that by increasing the connotation of a name we decrease its denotation: if, for example, to the definition of 'lion' we add 'inhabiting Africa,' Asiatic lions are no longer denoted by it. On the other hand, if we use a name to denote objects that it did not formerly apply to, some of the connotation must be dropped: if, for example, the name 'lion' be used to include 'pumas,' the tufted tail and mane can no longer be part of the meaning of the word; since pumas have not these properties.

    This doctrine is logically or formally true, but it may not always be true in fact. It is logically true; because wherever we add to the connotation of a name, it is possible that some things to which it formerly applied are now excluded from its denotation, though we may not know of any such things. Still, as a matter of fact, an object may be discovered to have a property previously unknown, and this property may be fundamental and co-extensive with the denotation of its name, or even more widely prevalent. The discovery that the whale is a mammal did not limit the class 'whale'; nor did the discovery that lions, dogs, wolves, *etc.*, walk upon their toes, affect the application of any of these names.

Similarly, the extension of a name to things not previously denoted by it, may not in fact alter its definition; for the extension may be made on the very ground that the things now first denoted by it have been found to have the properties enumerated in its definition, as when the name 'mammal' was applied to whales, dolphins, *etc.* If, however, 'mammal' had formerly been understood to apply only to land animals, so that its definition included (at least, popularly) the quality of 'living on the land,' this part of the connotation was of course lost when the denotation came to include certain aquatic animals.

A proprium is an attribute derived from the definition: being either (*a*) implied in it, or deducible from it, as 'having its three angles equal to two right angles' may be proved from the definition of a triangle; or (*b*) causally dependent on it, as being 'dangerous to flocks' results from the nature of a wolf, and as 'moving in an ellipse' results from the nature of a planet in its relation to the sun.

An accident is a property accompanying the defining attributes without being deducible from them. The word suggests that such a property is merely 'accidental,' or there 'by chance'; but it only means that we do not understand the connection.

Proprium and Accident bear the same relation to one another as Derivative and Empirical Laws: the predication of a proprium is a derivative law, and the predication of an accident is an empirical law. Both accidents and empirical laws present problems, the solution of which consists in reducing them, respectively, to propria and derivative laws. Thus the colour of animals was once regarded as an accident for which no reason could be given; but now the colour of animals is regarded as an effect of their nature and habits, the chief determinants of it being the advantage of concealment; whilst in other cases, as among brightly coloured insects and snakes, the determinant may be the advantage of advertising their own noxiousness. If such reasoning is sound, colour is a proprium (and if so, it cannot *logically* be included in a definition; but it is better to be judicious than formal).

If the colour of animals is a proprium, we must recognise a distinction between Inseparable and Separable Propria, according as they do, or do not, always accompany the essence: for mankind is regarded as one species; but each colour, white, black or yellow, is separable from it under different climatic conditions; whilst tigers are everywhere coloured and striped in much the same way; so that we may consider their colouring as inseparable, in spite of exceptional specimens black or white or clouded.

The same distinction may be drawn between accidents. 'Inhabiting Asia' is an Inseparable Accident of tiger, but a Separable Accident of lion. Even the occasional characteristics and occupations of individuals are sometimes called separable accidents of the species; as, of man, being colour-blind, carpentering, or running.

A proprium in the original signification of the term ἴδιον was peculiar to a species, never found with any other, and was therefore convertible with the subject; but this restriction is no longer insisted on.

§ 9. Any predication of a genus, difference or definition, is a verbal, analytic, or essential proposition: and any predication of a proprium or accident, is a real, synthetic, or accidental proposition (chap. v. § 6). A proposition is called verbal or analytic when the predicate is a part, or the whole, of the meaning of the subject; and the subject being species, a genus or difference is part, and a definition is the whole, of its meaning or connotation. Hence such a proposition has also been called explicative. Again, a proposition is called real or synthetic when the predicate is no part of the meaning of the subject; and, the subject being species, a proprium or accident is no part of its meaning or connotation. Hence such a proposition has been called ampliative.

As to Essential and Accidental, these terms are derived from the doctrine of Realism. Realists maintain that the essence of a thing, or that which makes a thing to be what (or of what kind) it is, also makes everything else of the same kind to be what it is. The essence, they say, is not proper to each thing or separately inherent in it, but is an 'Universal' common to all things of that kind. Some hold that the universal nature of things of any kind is an Idea existing (apart from the things) in the intelligible world, invisible to mortal eye and only accessible to thought; whence the Idea is called a noumenon: that only the Idea is truly real, and that the things (say, trees, bedsteads and cities) which appear to us in sense-perception, and which therefore are called phenomena, only exist by participating in, or imitating, the Idea of each kind of them. The standard of this school bears the legend *Universalia ante rem*.

But others think that the Universal does not exist apart from particular things, but is their present essence; gives them actuality as individual substances; "informs" them, or is their formal cause, and thus makes them to be what they are of their kind according to the definition: the universal lion is in all lions, and is not merely similar, but identical in all; for thus the Universal Reason thinks and energises in Nature. This school inscribes upon its banners, *Universalia in re*.

To define anything, then, is to discover its essence, whether transcendent or immanent; and to predicate the definition, or any part of it (genus or difference), is to enounce an essential proposition. But a proprium, being no part of a definition, though it always goes along with it, does not show what a thing is; nor of course does an accident; so that to predicate either of these is to enounce an accidental proposition.

Another school of Metaphysicians denies the existence of Universal Ideas or Forms; the real things, according to them, are individuals; which, so far as any of them resemble one another, are regarded as forming classes; and the only Universal is the class-name, which is applied universally in the same sense. Hence, they are called Nominalists. The sense in which any name is applied, they say, is derived from a comparison of the individuals, and by abstraction of the properties they have in common; and thus the definition is formed. *Universalia post rem* is their motto. Some Nominalists, however, hold that, though Universals do not exist in nature, they do in our minds, as Abstract Ideas or Concepts; and that to define a term is to analyse the concept it stands for; whence, these philosophers are called Conceptualists.

Such questions belong to Metaphysics rather than to Logic; and the foregoing is a commonplace account of a subject upon every point of which there is much difference of opinion.

§ 10. The doctrine of the Predicaments, or Categories, is so interwoven with the history of speculation and especially of Logic that, though its vitality is exhausted, it can hardly be passed over unmentioned. The predicaments of Aristotle are the heads of a classification of terms as possible predicates of a particular thing or individual. Hamilton (*Logic*: Lect. xi.) has given a classification of them; which, if it cannot be found in Aristotle, is an aid to the memory, and may be thrown into a table thus:

Substance	οὐσία	(1)

[Quantity	ποσόν	(2)
[Quality	ποιόν	(3)
[Relation	πρός τι	(4)
[Where	ποῦ	(5)
[When	πότε	(6)
[Action	ποιεῖν	(7)
[Passion	πάσχειν	(8)
[Posture	κεῖσθαι	(9)
[Habit	ἔχειν	(10)

Taking a particular thing or individual, as 'Socrates,' this is Substance in the proper sense of the word, and can never be a predicate, but is the subject of all predicates. We may assert of him (1) Substance in the secondary sense (species or genus) that he is a man or an animal; (2) Quantity, of such a height or weight; (3) Quality, fair or dark; (4) Relation, shorter or taller than Xanthippe; (5) Where, at Athens; (6) When, two thousand and odd years ago; (7) Action, that he questions or pleads; (8) Passion, that he is answered or condemned; (9) Posture, that he sits or stands; (10) Habit, that he is clothed or armed.

Thus illustrated (*Categoriæ*: c. 4), the predicaments seem to be a list of topics, generally useful for the analysis and description of an individual, but wanting in the scientific qualities of rational arrangement, derivation and limitation. Why are there just these heads, and just so many? It has been suggested that they were determined by grammatical forms: for Substance is expressed by a substantive; Quantity, Quality and Relation are adjectival; Where and When, adverbial; and the remaining four are verbal. It is true that the parts of speech were not systematically discriminated until some years after Aristotle's time; but, as they existed, they may have unconsciously influenced his selection and arrangement of the predicaments. Where a principle is so obscure one feels glad of any clue to it (*cf.* Grote's *Aristotle*, c. 3, and Zeller's *Aristotle*, c. 6). But whatever the origin and original meaning of the predicaments, they were for a long time regarded as a classification of things; and it is in this sense that Mill criticises them (*Logic*: Bk. I. c. 3).

If, however, the predicaments are heads of a classification of terms predicable, we may expect to find some connection with the predicables; and, in fact, secondary Substances are species and genus; whilst the remaining nine forms are generally accidents. But, again, we may expect some agreement between them and the fundamental forms of predication (*ante*, chap. i. § 5, and chap. ii § 4): Substance, whether as the foundation of attributes, or as genus and species, implies the predication of co-inherence, which is one mode of *Co-existence*. Quantity is predicated as equality (or inequality) a mode of *Likeness*; and the other mode of *Likeness* is involved in the predication of Quality. Relation, indeed, is the abstract of all predication, and ought not to appear in a list along with special forms of itself. 'Where' is position, or *Co-existence* in space; and 'When' is position in time, or *Succession*. Action and Passion are the most interesting aspect of *Causation*. Posture and Habit are complex modes of *Co-existence*, but too specialised to have any philosophic value. Now, I do not pretend that this is what Aristotle meant and was trying to say: but if Likeness, Co-existence, Succession and Causation are fundamental forms of predication, a good mind analysing the fact of predication is likely to happen upon them in one set of words or another.

By Kant the word 'Category' has been appropriated to the highest forms of judgment, such as Unity, Reality, Substance, and Cause, under which the understanding reduces phenomena to order and thereby constitutes Nature. This change of meaning has not been made without a certain continuity of thought; for forms of judgment are modes of predication. But besides altering the lists of categories and greatly improving it, Kant has brought forward under an old title a doctrine so original and suggestive that it has extensively influenced the subsequent history of Philosophy. At the same time, and probably as a result of the vogue of the Kantian philosophy, the word 'category' has been vulgarised as a synonym for 'class,' just as 'predicament' long ago passed from Scholastic Logic into common use as a synonym for 'plight.' A minister is said to be 'in a predicament,' or to fall under the 'category of impostors.'

CHAPTER XXIII

DEFINITION OF COMMON TERMS

§ 1. Ordinary words may need definition, if in the course of exposition or argument their meaning is liable to be mistaken. But as definition cannot give one the sense of a popular word for all occasions of its use, it is an operation of great delicacy. Fixity of meaning in the use of single words is contrary to the genius of the common vocabulary; since each word, whilst having a certain predominant character, must be used with many shades of significance, in order to express the different thoughts and feelings of multitudes of men in endlessly diversified situations; and its force, whenever it is used, is qualified by the other words with which it is connected in a sentence, by its place in the construction of the sentence, by the emphasis, or by the pitch of its pronunciation compared with the other words.

Clearly, the requisite of a scientific language, 'that every word shall have one meaning well defined,' is too exacting for popular language; because the other chief requisite of scientific language cannot be complied with, 'that there be no important meaning without a name.' 'Important meanings,' or what seem such, are too numerous to be thus provided for; and new ones are constantly arising, as each of us pursues his business or his pleasure, his meditations or the excursions of his fancy. It is impossible to have a separate term for each meaning; and, therefore, the terms we have must admit of variable application.

An attempt to introduce new words is generally disgusting. Few men have mastered the uses of half the words already to be found in our classics. Much more would be lost than gained by doubling the dictionary. It is true that, at certain stages in the growth of a people, a need may be widely felt for the adoption of new words: such, in our own case, was the period of the Tudors and early Stuarts. Many fresh words, chiefly from the Latin, then appeared in books, were often received with reprobation and derision, sometimes disappeared again, sometimes established their footing in the language: see *The Art of English Poetry* (ascribed to Puttenham), Book III. chap. 4, and Ben Jonson's *Poetaster*, Act. V. sc. I. Good judges did not know whether a word was really called for: even Shakespeare thought 'remuneration' and 'accommodate' ridiculous. But such national exigencies rarely arise; and in our own time great authors distinguish themselves by the plastic power with which they make common words convey uncommon meanings.

Fluid, however, as popular language is and ought to be, it may be necessary for the sake of clear exposition, or to steady the course of an argument, to avoid either sophistry or unintentional confusion, that words should be defined and discriminated; and we must discuss the means of doing so.

§ 2. Scientific method is applicable, with some qualifications, to the definition of ordinary words. Classification is involved in any problem of definition: at least, if our object is to find a meaning that shall be generally acceptable and intelligible. No doubt two disputants may, for their own satisfaction, adopt any arbitrary definition of a word important in their controversy; or, any one may define a word as he pleases, at the risk of being misunderstood, provided he has no fraudulent intention. But in exposition or argument addressed to the public, where words are used in some of their ordinary senses, it should be recognised that the meaning of each one involves that of many others. For language has grown with the human mind, as representing its knowledge of the world: this knowledge consists of the resemblances and differences of things and of the activities of things, that is, of classes and causes; and as there is such order in the world, so there must be in language: language, therefore, embodies an irregular classification of things with their attributes and relations according to our knowledge and beliefs. The best attempt (known to me) to carry out this view is contained in Roget's *Thesaurus*, which is a classification of English words according to their meanings: founded, as the author tells us, on the models of Zoology and Botany, it has some of the requisites of a Logical Dictionary.

Popular language, indeed, having grown up with a predominantly practical purpose, represents a very imperfect classification philosophically considered. Things, or aspects, or processes of things, that have excited little interest, have often gone unnamed: so that scientific discoverers are obliged, for scientific purposes, to invent thousands of new names. Strong interests, on the other hand, give such a colour to language, that, where they enter, it is difficult to find any indifferent expressions. *Consistency* being much prized, though often the part of a blockhead, *inconsistency* implies not merely the absence of the supposed virtue, but a positive vice: *Beauty* being attractive and *ugliness* the reverse, if we invent a word for that which is neither, 'plainness,' it at once becomes tinged with the ugly. We seem to love beauty and morality so much as to be almost incapable of signifying their absence without expressing aversion.

Again, the erroneous theories of mankind have often found their way into popular speech, and their terms have remained there long after the rejection of the beliefs they embodied: as—lunatic, augury, divination, spell, exorcism: though, to be sure, such words may often be turned to good account, besides the interest of preserving their original sense. Language is a record as well as an index of ideas.

Language, then, being essentially classificatory, any attempt to ascertain the meaning of a word, far from neglecting its relations to others, should be directed toward elucidating them.

Every word belongs to a group, and this group to some other larger group. A group is sometimes formed by derivation, at least so far as different meanings are marked merely by inflections, as *short, shorter, shorten, shortly*; but, for the most part, is a conflux of words from many different sources. *Repose, depose, suppose, impose, propose*, are not nearly connected in meaning; but are severally allied in sense much more closely with words philologically remote. Thus *repose* is allied with *rest, sleep, tranquillity; disturbance, unrest, tumult*; whilst *depose* is, in one sense, allied with *overthrow, dismiss, dethrone; restore, confirm, establish*; and, in another sense, with *declare, attest, swear, prove, etc.* Groups of words, in fact, depend on their meanings, just as the connection of scientific names follows the resemblance in character of the things denoted.

Words, accordingly, stand related to one another, for the most part, though very irregularly, as genus, species, and co-ordinate species. Taking *repose* as a genus, we have as species of it, though not exactly co-ordinate with one another, *tranquillity* with a mental differentia (repose of mind), *rest*, whether of mind or body, *sleep*, with the differentia of unconsciousness (privative). Synonyms are species, or varieties, wherever any difference can be detected in them; and to discriminate them we must first find the generic meaning; for which there may, or may not, be a single word. Thus, *equality, sameness, likeness, similarity, resemblance, identity*, are synonyms; but, if we attend to the ways in which they are actually used, perhaps none of them can claim to be a genus in relation to the rest. If so, we must resort to a compound term for the genus, such as 'absence of some sort of difference.' Then *equality* is absence of difference in quantity; *sameness* is often absence of difference in quality, though the usage is not strict: *likeness, similarity*, and *resemblance*, in their actual use, perhaps, cannot be discriminated; unless *likeness* be the more concrete, *similarity* the more abstract; but they may all be used compatibly with the recognition of more or less difference in the things compared, and even imply this. *Identity* is the absence of difference of origin, a continuity of existence, with so much sameness from moment to moment as is compatible with changes in the course of nature; so that egg, caterpillar, chrysalis, butterfly may be identical for the run of an individual life, in spite of differences quantitative and qualitative, as truly as a shilling that all the time lies in a drawer.

Co-ordinate Species, when positive, have the least contrariety; but there are also opposites, namely, negatives, contradictories and fuller contraries. These may be regarded as either co-ordinate genera or the species of co-ordinate genera. Thus, *repose* being a genus, *not-repose* is by dichotomy a co-ordinate genus and is a negative and contradictory; then *activity* (implying an end in view), *motion* (limited to matter), *disturbance* (implying changes from a state of calm), *tumult*, etc., are co-ordinate species of *not-repose*, and are therefore co-ordinate opposites, or contraries, of the species of *repose*.

As for correlative words, like *master and slave, husband and wife*, etc., it may seem far-fetched to compare them with the sexes of the same species of plants or animals; but there is this resemblance between the two cases, that sexual names are correlative, as 'lioness,' and that one sex of a species, like a correlative name, cannot be defined without implying the other; for if a distinctive attribute of one sex be mentioned (as the lion's mane), it is implied that the other wants it, and apart from this implication the species is not defined: just as the definition of 'master' implies a 'slave' to obey.

Common words, less precise than the terms of a scientific nomenclature, differ from them also in this, that the same word may occur in different genera. Thus, *sleep* is a species of *repose* as above; but it is also a species of *unconsciousness*, with co-ordinate species *swoon, hypnotic state, etc.* In fact, every word stands under as many distinct genera, at least, as there are simple or indefinable qualities to be enumerated in its definition.

§ 3. Partially similar to a scientific nomenclature, ordinary language has likewise a terminology for describing things according to their qualities and structure. Such is the function of all the names of colours, sounds, tastes, contrasts of temperature, of hardness, of pleasantness; in short, of all descriptive adjectives, and all names for the parts and processes of things. Any word connoting a quality may be used to describe many very different things, as long as they agree in that quality.

But the quality connoted by a word, and treated as always the same quality, is often only analogically the same. We speak of a *great* storm, a *great* man, a *great* book; but *great* is in each case not only relative, implying small, and leaving open the possibility that what we call great is still smaller than something else of its kind, but it is also predicated with reference to some quality or qualities, which may be very different in the several cases of its application. If the book is prized for wisdom, or for imagination, its greatness lies in that quality; if the man is distinguished for influence, or for courage, his greatness is of that nature; if the storm is remarkable for violence, or for duration, its greatness depends on that fact. The word *great*, therefore, is not used for these things in the same sense, but only analogically and elliptically. Similarly with good, pure, free, strong, rich, and so on. 'Rest' has not the same meaning in respect of a stone and of an animal, nor 'strong' in respect of thought and muscle, nor 'sweet' in respect of sugar and music. But here we come to the border between literal and figurative use; every one sees that figurative epithets are analogical; but by custom any figurative use may become literal.

Again, many general names of widely different meaning, are brought together in describing any concrete object, as an animal, or a landscape, or in defining any specific term. This is the sense of the doctrine, that any concrete thing is a conflux of generalities or universals: it may at least be considered in this way; though it seems more natural to say, that an object presents these different aspects to a spectator, who, fully to comprehend it, must classify it in every aspect.

§ 4. The process of seeking a definition may be guided by the following maxims:

(1) Find the usage of good modern authors; that is (as they rarely define a word explicitly), consider what in various relations they use it to denote; from which uses its connotation may be collected.

(2) But if this process yield no satisfactory result, make a list of the things denoted, and of those denoted by the co-ordinate and opposite words; and observe the qualities in which the things denoted agree, and in which they differ from those denoted by the contraries and opposites. If 'civilisation' is to be defined, make lists of civilised peoples, of semi-civilised, of barbarous, and of savage: now, what things are common to civilised peoples and wanting in the others respectively? This is an exercise worth attempting. If poetry is to be defined, survey some typical examples of what good critics recognise as poetry, and compare them with examples of bad 'poetry,' literary prose, oratory, and science. Having determined the characteristics of each kind, arrange them opposite one another in parallel columns. Whoever tries to define by this method a few important, frequently occurring words, will find his thoughts the clearer for it, and will collect by the way much information which may be more valuable than the definition itself, should he ever find one.

(3) If the genus of a word to be defined is already known, the process may be shortened. Suppose the genus of poetry to be *belles lettres* (that is, 'appealing to good taste'), this suffices to mark it off from science; but since literary prose and oratory are also *belles lettres*, we must still seek the differentia of poetry by a comparison of it with these co-ordinate species. A compound word often exhibits genus and difference upon its face: as 're-turn,' 'inter-penetrate,' 'tuning-fork,' 'cricket-bat'; but the two last would hardly be understood without inspection or further description. And however a definition be discovered, it is well to state it *per genus et differentiam*.

(4) In defining any term we should avoid encroaching upon the meaning of any of the co-ordinate terms; for else their usefulness is lessened: as by making 'law' include 'custom,' or 'wealth' include 'labour' or 'culture.'

(5) If two or more terms happen to be exactly synonymous, it may be possible (and, if so, it is a service to the language) to divert one of them to any neighbouring meaning that has no determinate expression. Thus, Wordsworth and Coleridge took great pains to distinguish between Imagination and Fancy, which had become in common usage practically equivalent; and they sought to limit 'imagination' to an order of poetic effect, which (they said) had prevailed during the Elizabethan age, but had been almost lost during the Gallo-classic, and which it was their mission to restore. Co-ordinate terms often tend to coalesce and become synonymous, or one almost supersedes the other, to the consequent impoverishment of our speech. At present *proposition* (that something is the fact) has almost driven out *proposal* (that it is desirable to co-operate in some action). Even good writers and speakers, by their own practice, encourage this confusion: they submit to Parliament certain 'propositions' (proposals for legislation), or even make 'a proposition of marriage.' Definition should counteract such a tendency.

(6) We must avoid the temptation to extend the denotation of a word so far as to diminish or destroy its connotation; or to increase its connotation so much as to render it no longer applicable to things which it formerly denoted: we should neither unduly generalise, nor unduly specialise, a term. Is it desirable to define *education* so as to include the 'lessons of experience'; or is it better to restrict it as implying a personal educator? If any word implies blame or praise, we are apt to extend it to everything we hate or approve. But *coward* cannot be so defined as to include all bullies, nor *noble* so as to include every honest man, without some loss in distinctness of thought.

The same impulses make us specialise words; for, if two words express approval, we wish to apply both to whatever we admire and to refuse both to whatever displeases us. Thus, a man may resolve to call no one great who is not good: greatness, according to him, connotes goodness: whence it follows that (say) Napoleon I. was not great. Another man is disgusted with greatness: according to him, good and great are mutually exclusive classes, sheep and goats, as in Gray's wretched clench: "Beneath the good how far, yet far above the great." In feet, however 'good' and 'great' are descriptive terms, sometimes applicable to the same object, sometimes to different: but 'great' is the wider term and applicable to goodness itself and also to badness; whereas by making 'great' connote goodness it becomes the narrower term. And as we have seen (§ 3), such epithets may be applicable to objects on account of different qualities: *good* is not predicated on the same ground of a man and of a horse.

(7) In defining any word, it is desirable to bear in mind its derivation, and to preserve the connection of meaning with its origin; unless there are preponderant reasons for diverting it, grounded on our need of the word to express a certain sense, and the greater difficulty of finding any other word for the same purpose. It is better to lean to the classical than to the vulgar sense of 'indifferent,' 'impertinent,' 'aggravating,' 'phenomenal.'

(8) Rigorous definition should not be attempted where the subject does not admit of it. Some kinds of things are so complex in their qualities, and each quality may manifest itself in so many degrees without ever admitting of exact measurement, that we have no means of marking them off precisely from other things nearly allied, similarly complex and similarly variable. If so we cannot precisely define their names. Imagination and fancy are of this nature, civilisation and barbarism, poetry and other kinds of literary expression. As to poetry, some think it only exists in metre, but hardly maintain that the metre must be strictly regular: if not, how much irregularity of rhythm is admissible? Others regard a certain mood of impassioned imagination as the essence of poetry; but they have never told us how great intensity of this mood is requisite. We also hear that poetry is of such a nature that the enjoyment of it is an end in itself; but as it is not maintained that poetry must be wholly impersuasive or uninstructive, there seems to be no means of deciding what amount or prominence of persuasion or instruction would transfer the work to the region of oratory or science. Such cases make the method of defining by the aid of a type really useful: the difficulty can hardly be got over without pointing to typical examples of each meaning, and admitting that there may be many divergences and unclassifiable instances on the border between allied meanings.

§ 5. As science began from common knowledge, the terms of the common vocabulary have often been adopted into the sciences, and many are still found there: such as weight, mass, work, attraction, repulsion, diffusion, reflection, absorption, base, salt, and so forth. In the more exact sciences, the vague popular associations with such words are hardly an inconvenience: since those addicted to such studies do not expect to master them without undergoing special discipline; and, having precisely defined the terms, they acquire the habit of thinking with them according to their assigned signification in those investigations to which they are appropriate. It is in the Social Sciences, especially Economics and Ethics, that the use of popular terminology is at once unavoidable and prejudicial. For the subject-matters, industry and the conduct of life, are every man's business; and, accordingly, have always been discussed with a consciousness of their direct practical bearing upon public and private interests, and therefore in the common language, in order that everybody may as far as possible benefit by whatever light can be thrown upon them. The general practice of Economists and Moralists, however, shows that, in their judgment, the good derived from writing in the common vocabulary outweighs the evil: though it is sometimes manifest that they themselves have been misled by extra-scientific meanings. To reduce the evil as much as possible, the following precautions seem reasonable:

(1) To try to find and adopt the central meaning of the word (say rent or money) in its current or traditional applications: so as to lessen in the greater number of cases the jar of conflicting associations. But if the central popular meaning does not correspond with the scientific conception to be expressed, it may be better to invent a new term.

(2) To define the term with sufficient accuracy to secure its clear and consistent use for scientific purposes.

(3) When a popular term has to be used in a sense that departs from the ordinary one in such a way as to incur the danger of misunderstanding, to qualify it by some adjunct or "interpretation-clause."

The first of these rules is not always adhered to; and, in the progress of a science, as subtler and more abstract relations are discovered amongst the facts, the meaning of a term may have to be modified and shifted further and further from its popular use. The term 'rent,' for example, is used by economists, in such a sense that they have to begin the discussion of it by explaining that it does not imply any actual payment by one man to another. Here, for most readers, the meaning they are accustomed to, seems already to have entirely disappeared. Difficulties may, however, be largely overcome by qualifying the term in its various relations, as produce-rents, ground-rents, customary rents, and so forth, (*Cf.* Dr. Keynes' *Scope and Method of Political Economy*, chap. 5.)

§ 6. Definitions affect the cogency of arguments in many ways, whether we use popular or scientific language. If the definitions of our terms are vague, or are badly abstracted from the facts denoted, all arguments involving these terms are inconclusive. There can be no confidence in reasoning with such terms; since, if vague, there is nothing to protect us from ambiguity; or, if their meaning has been badly abstracted, we may be led into absurdity—as if 'impudence' should be defined in such a way as to confound it with honesty.

Again, it is by definitions that we can best distinguish between Verbal and Real Propositions. Whether a term predicated is implied in the definition of the subject, or adds something to its meaning, deserves our constant attention. We often persuade ourselves that statements are profound and important, when, in fact, they are mere verbal propositions. "It is just to give every man his due"; "the greater good ought to be preferred to the less"; such dicta sound well—indeed, too well! For 'a man's due' means nothing else than what it is just to give him; and 'the greater good' may mean the one that ought to be preferred. The investigation of a definition may be a very valuable service to thought; but, once found, there is no merit in repeating it. To put forward verbal or analytic propositions, or truisms, as information (except, of course, in explaining terms to the uninstructed), shows that we are not thinking what we say; for else we must become aware of our own emptiness. Every step forward in knowledge is expressed in a real or synthetic proposition; and it is only by means of such propositions that information can be given (except as to the meaning of words) or that an argument or train of reasoning can make any progress.

Opposed to a truism is a Contradiction in Terms; that is, the denying of a subject something which it connotes (or which belongs to its definition), or the affirming of it something whose absence it connotes (or which is excluded by its definition). A verbal proposition is necessarily true, because it is tautologous; a contradiction in terms is necessarily false, because it is inconsistent. Yet, as a rhetorical artifice, or figure, it may be effective: that 'the slave is not bound to obey his master' may be a way of saying that there ought to be no slaves; that 'property is theft,' is an uncompromising assertion of the communistic ideal. Similarly a truism may have rhetorical value: that 'a Negro is a man' has often been a timely reminder, or even that "a man's a man." It is only when we fall into such contradiction or tautology by lapse of thought, by not fully understanding our own words, that it becomes absurd.

Real Propositions comprise the predication of Propria and Accidentia. Accidentia, implying a sort of empirical law, can only be established by direct induction. But propria are deduced from (or rather by means of) the definition with the help of real propositions, and this is what is called 'arguing from a Definition.' Thus, if increasing capacity for co-operation be a specific character of civilisation, 'great wealth' may be considered as a proprium of civilised as compared with barbarous nations. For co-operation is made most effectual by the division of labour, and that this is the chief condition of producing wealth is a real proposition. Such arguments from definitions concerning concrete facts and causation require verification by comparing the conclusion with the facts. The verification of this example is easy, if we do not let ourselves be misled in estimating the wealth of barbarians by the ostentatious "pearl and gold" of kings and nobles, where 99 per cent. of the people live in penury and servitude. The wealth of civilisation is not only great but diffused, and in its diffusion its greatness must be estimated.

To argue from a definition may be a process of several degrees of complexity. The simplest case is the establishing of a proprium as the direct consequence of some connoted attribute, as in the above example. If the definition has been correctly abstracted from the particulars, the particulars have the attributes summarised in the definition; and, therefore, they have whatever can be shown to follow from those attributes. But it frequently happens that the argument rests partly on the qualities connoted by the class name and partly on many other facts.

In Geometry, the proof of a theorem depends not only upon the definition of the figure or figures directly concerned, but also upon one or more axioms, and upon propria or constructions already established. Thus, in Euclid's fifth Proposition, the proof that the angles at the base of an isosceles triangle are equal, depends not only on the equality of the opposite sides, but upon this together with the construction that shows how from the greater of two lines a part may be cut off equal to the less, the proof that triangles that can be conceived to coincide are equal, and the axiom that if equals be taken from equals the remainders are equal. Similarly, in Biology, if colouring favourable to concealment is a proprium of carnivorous animals, it is not deducible merely from their predatory character or any other attribute entering into the definition of any species of them, but from their predatory character together with the causes summarised in the phrase 'Natural Selection'; that is, competition for a livelihood, and the destruction of those that labour under any disadvantages, of which conspicuous colouring would be one. The particular coloration of any given species, again, can only be deduced by further considering its habitat (desert, jungle or snowfield): a circumstance lying wholly outside the definition of the species.

The validity of an argument based partly or wholly on a definition depends, in the first place, on the existence of things corresponding with the definition—that is, having the properties connoted by the name defined. If there are no such things as isosceles triangles, Euclid's fifth Proposition is only formally true, like a theorem concerning the fourth dimension of space: merely consistent with his other assumptions. But if there be any triangles only approximately isosceles, the proof applies to them, making allowance for their concrete imperfection: the nearer their sides approach straightness and equality the more nearly equal will the opposite angles be.

Again, as to the things corresponding with terms defined, according to Dr. Venn, their 'existence' may be understood in several senses: (1) merely for the reason, like the pure genera and species of Porphyry's tree; the sole condition of whose being is logical consistency: or (2) for the imagination, like the giants and magicians of romance, the heroes of tragedy and the fairies of popular superstition; whose properties may be discussed, and verified by appeal to the right documents and authorities (poems and ballads): or (3) for perception, like plants, animals, stones and stars. Only the third class exist in the proper sense of the word. But under a convention or hypothesis of existence, we may argue from the definition of a fairy, or a demigod, or a dragon, and deduce various consequences without absurdity, if we are content with poetic consistency and the authority of myths and romances as the test of truth.

In the region of concrete objects, whose properties are causes, and neither merely fictions nor determinations of space (as in Geometry), we meet with another condition of the validity of any argument depending on a definition: there must not only be objects corresponding to the definition, but there must be no other causes counteracting those qualities on whose agency our argument relies. Thus, though we may infer from the quality of co-operation connoted by civilisation, that a civilised country will be a wealthy one, this may not be found true of such a country recently devastated by war or other calamity. Nor can co-operation always triumph over disadvantageous circumstances. Scandinavia is so poor in the gifts of nature favourable to industry, that it is not wealthy in spite of civilisation: still, it is far wealthier than it would be in the hands of a barbarous people. In short, when arguing from a definition, we can only infer the *tendency* of any causal characteristics included in it; the unqualified realisation of such a tendency must depend upon the absence of counteracting causes. As soon as we leave the region of pure conceptions and make any attempt to bring our speculations home to the actual phenomena of nature or of human life, the verification of every inference becomes an unremitting obligation.

CHAPTER XXIV

FALLACIES

§ 1. A Fallacy is any failure to fulfil the conditions of proof. If we neglect or mistake the conditions of proof unintentionally, whether in our private meditations or in addressing others, it is a Paralogism: but if we endeavour to pass off upon others evidence or argument which we know or suspect to be unsound, it is a Sophism.

Fallacies, whether paralogisms or sophisms, may be divided into two classes: (*a*) the Formal, or those that can be shown to conflict with one or more of the truths of Logic, whether Deductive or Inductive; as if we attempt to prove an universal affirmative in the Third Figure; or to argue that, as the average expectation of life for males at the age of 20 is 19½ years, therefore Alcibiades, being 20 years of age, will die when he is 39½; (*b*) the Material, or those that cannot be clearly exhibited as transgressions of any logical principle, but are due to superficial inquiry or confused reasoning; as in adopting premises on insufficient authority, or without examining the facts; or in mistaking the point to be proved.

§ 2. Formal Fallacies of Deduction and Induction are, all of them, breaches of the rule 'not to go beyond the evidence.' As a detailed account of them would be little else than a repetition of the foregoing chapters, it may suffice to recall some of the places at which it is easiest to go astray.

(1) It is not uncommon to mistake the Contrary for the Contradictory, as—A is not taller than B, ∴ he is shorter.

(2) To convert *A.* or *O.* simply, as—
All Money is Wealth ∴ All Wealth is Money;
or—Some Wealth is not Money ∴ Some Money is not Wealth.
In both these cases, Wealth, though undistributed in the convertend, is distributed in the converse.

(3) To attempt to syllogise with two premises containing four terms, as
The Papuans are savages; The Javanese are neighbours of the Papuans: ∴ The Javanese are savages.

Such an argument is excluded by the definition of a Syllogism, and presents no formal evidence whatever. We should naturally assume that any man who advanced it merely meant to raise some probability that 'neighbourhood is a sign of community of ideas and customs.' But, if so, he should have been more explicit. There would, of course, be the same failure of connection, if a fourth term were introduced into the conclusion, instead of into the premises.

(4) To distribute in the conclusion a term that was undistributed in the premises (an error essentially the same as (2) above), *i.e.*, Illicit process of the major or minor term, as—
Every rational agent is accountable; Brutes are not rational agents: ∴ Brutes are not accountable.

In this example (from Whately), an illegitimate mood of Fig. I., the major term, 'accountable,' has suffered the illicit process; since, in the premise, it is predicate of an affirmative proposition and, therefore, undistributed; but, in the conclusion, it is predicate of a negative proposition and, therefore, distributed. The fact that nearly everybody would accept the conclusion as true, might lead one to overlook the formal inconclusiveness of the proof.

Again,
All men are two-handed; All two-handed animals are cooking animals: ∴ All cooking animals are men.

Here we have Bramantip concluding in A.; and there is, formally, an illicit process of the minor; though the conclusion is true; and the evidence, such as it is, is materially adequate. ('Two-handed,' being a peculiar differentia, is nugatory as a middle term, and may be cut out of both premises; whilst 'cooking' is a proprium peculiar to the species Man; so that these terms might be related in U., *All men are all cookers*; whence, by conversion, *All cookers are men.*)

(5) To omit to distribute the middle term in one or the other premise, as—
All verbal propositions are self-evident; All axioms are self-evident: ∴ All axioms are verbal propositions.

This is an illegitimate mood in Fig. II.; in which, to give any conclusion, one premise must be negative. It may serve as a formal illustration of Undistributed Middle; though, as both premises are verbal propositions, it is (materially) not syllogistic at all, but an error of classification; a confounding of co-ordinate species by assuming their identity because they have the generic attribute in common.

(6) To simply convert an hypothetical proposition, as—
If trade is free, it prospers; ∴ If trade prospers, it is free.

This is similar to the simple conversion of the categorical A.; since it takes for granted that the antecedent is co-extensive with the consequent, or (in other words) that the freedom of trade is the sole condition of, or (at least) inseparable from, its prosperity.

The same assumption is made if, in an hypothetical syllogism, we try to ground an inference on the affirmation of the consequent or denial of the antecedent, as—
If trade is free it prospers: It does prosper; ∴ It is free. It is not free; ∴ It does not prosper.

Neither of these arguments is formally good; nor, of course, is either of them materially valid, if it be possible for trade to prosper in spite of protective tariffs.

An important example of this fallacy is the prevalent notion, that if the conclusion of an argument is true the premises must be trustworthy; or, that if the premises are false the conclusion must be erroneous. For, plainly, that—
If the premises are true, the conclusion is true, is a hypothetical proposition; and we argue justly—
The premises are true; ∴ The conclusion is true; or, The conclusion is false; ∴ The premises are false (or one of them is).

This is valid for every argument that is formally correct; but that we cannot trust the premises on the strength of the conclusion, nor reject the conclusion because the premises are absurd, the following example will show:
All who square the circle are great mathematicians; Newton squared the circle: ∴ Newton was a great mathematician.
The conclusion is true; but the premises are intolerable.

How the taking of Contraries for Contradictories may vitiate Disjunctive Syllogisms and Dilemmas has been sufficiently explained in the twelfth chapter.

§ 3. Formal Fallacies of Induction consist in supposing or inferring Causation without attempting to prove it, or in pretending to prove it without satisfying the Canons of observation and experiment: as—

(1) To assign the Cause of anything that is not a concrete event: as, *e.g.*, why two circles can touch only in one point. We should give the 'reason'; for this expression includes, besides evidence of causation, the principles of formal deduction, logical and mathematical.

(2) To argue, as if on inductive grounds, concerning the cause of the Universe as a whole. This may be called the fallacy of transcendent inference: since the Canons are only applicable to instances of events that can be compared; they cannot deal with that which is in its nature unique.

(3) To mistake co-existent phenomena for cause and effect: as when a man, wearing an amulet and escaping shipwreck, regards the amulet as the cause of his escape. To prove his point, he must either get again into exactly the same circumstances without his amulet, and be drowned—according to the method of Difference; or, shirking the only satisfactory test, and putting up with mere Agreement, he must show, (*a*) that all who are shipwrecked and escape wear amulets, and (*b*) that their cases agree in nothing else; and (*c*), by the Joint Method, that all who are shipwrecked without amulets are drowned. And even if his evidence, according to Agreement, seemed satisfactory at all these points, it would still be fallacious to trust to it as proof of direct causation; since we have seen that unaided observation is never sufficient for this: it is only by experiment in prepared circumstances that we can confidently trace sequence and the transfer of energy.

There is the reverse error of mistaking causal connection for independent co-existence: as if any one regards it as merely a curious coincidence that great rivers generally flow past great towns. In this case, however, the evidence of connection does not depend merely upon direct Induction.

(4) *Post hoc, ergo propter hoc*: to accept the mere sequence of phenomena, even though often repeated, as proving that the phenomena are cause and effect, or connected by causation. This is a very natural error: for although, the antecedents of a phenomenon being numerous, most of them cannot be its cause, yet it is among them that the cause must be sought. Indeed, if there is neither time nor opportunity for analysis, it may seem better to accept any antecedent as a cause (or, at least, as a sign) of an important event than to go without any guide. And, accordingly, the vast and complicated learning of omens, augury, horoscopy and prophetic dreams, relies upon this maxim; for whatever the origin of such superstitions, a single coincidence in their favour triumphantly confirms them. It is the besetting delusion of everybody who has wishes or prejudices; that is, of all of us at some time or other; for then we are ready to believe without evidence. The fallacy consists in judging off-hand, without any attempt, either by logic or by common sense, to eliminate the irrelevant antecedents; which may include all the most striking and specious.

(5) To regard the Co-Effects (whether simultaneous or successive) of a common cause as standing in the direct relation of cause and effect. Probably no one supposes that the falling of the mercury in his thermometer causes the neighbouring lake to freeze. True, it is the antecedent, and (within a narrow range of experience) may be the invariable antecedent of the formation of ice; but, besides that the two events are so unequal, every one is aware that there is another antecedent, the fall of temperature, which causes both. To justify inductively our belief in causation, the instances compared must agree, or differ, in one circumstance only (besides the effect). The flowing tide is an antecedent of the ebbing tide; it is invariably so, and is equal to it; but it is not the cause of it: other circumstances are present; and the moon is the chief condition of both flow and ebb. In several instances, States that have grown outrageously luxurious have declined in power: that luxury caused their downfall may seem obvious, and capable of furnishing a moral lesson to the young. Hence other important circumstances are overlooked, such as the institution of slavery, the corruption and rapacity of officials and tax-gatherers, an army too powerful for discipline; any or all of which may be present, and sufficient to explain both the luxury and the ruin.

(6) To mistake one condition of a phenomenon for the whole cause. To speak of an indispensable condition of any phenomenon as the cause of it, may be a mere conventional abbreviation; and in this way such a mode of expression is common not only in popular but also in scientific discussion. Thus we say that a temperature of 33° F. is a cause of the melting of ice; although that ice melts at 33° F., must further depend upon something in the nature of water; for every solid has its own melting-point. As long, then, as we remember that 'cause,' used in this sense, is only a convenient abbreviation, no harm is done; but, if we forget it, fallacy may result: as when a man says that the cause of a financial crisis was the raising of the rate of discount, neglecting the other conditions of the market; whereas, in some circumstances, a rise of the Bank-rate may increase public confidence and prevent a crisis.

We have seen that the direct use of the Canons of Agreement and Difference may only enable us to say that a certain antecedent is a cause or an indispensable condition of the phenomenon under investigation. If, therefore, it is important to find the whole cause, we must either experiment directly upon the other conditions, or resort to the Method of Residues and deductive reasoning; nor must we be content, without showing (where such precision is possible) that the alleged cause and the given phenomenon are equal.

(7) To mistake a single consequence of a given cause for the whole effect, is a corresponding error; and none so common. Nearly all the mistakes of private conduct and of legislation are due to it: To cure temporary lassitude by a stimulant, and so derange the liver; to establish a new industry by protective duties, and thereby impoverish the rest of the country; to gag the press, and so drive the discontented into conspiracy; to build an alms-house, and thereby attract paupers into the parish, raise the rates, and discourage industry.

(8) To demand greater exactness in the estimate of causes or effects than a given subject admits of. In the more complex sciences, Biology, Psychology, Sociology, it is often impossible to be confident that all the conditions of a given phenomenon have been assigned, or that all its consequences have been traced. The causes of the origin of species and of the great French Revolution have been carefully investigated, and still we may doubt whether they have all been discovered, or whether their comparative importance has been rightly determined; but it would be very unreasonable to treat those things as miraculous and unintelligible. We read in the *Ethics*, that a properly cultivated mind knows what degree of precision is to be expected in each science. The greatest possible precision is always to be sought; but what is possible depends partly on the nature of the study and partly upon the state of scientific preparation.

(9) To treat an agent or condition remote in time as an unconditional cause: for every moment of time gives an opportunity for new combinations of forces and, therefore, for modifications of the effect. Thus, although we often say that Napoleon's Russian expedition was the cause of his downfall, yet the effect was subject to numerous further conditions. Had the natives not burnt Moscow, had the winter been exceptionally mild, had the Prussians and Austrians not risen against him, the event might have been very different. It is rash to trace the liberties of modern Europe to the battle of Marathon. Indeed, our powers of perception are so unequal to the subtlety of nature, that even in experimental science there is time for molecular changes to occur between what we treat as a cause and the effect as we perceive it; and, in such cases, the strictly unconditional cause has not been discovered.

(10) To neglect the negative conditions to which a cause is subject. When we say that water boils at 212° F., we mean "provided the pressure be the same as that of the atmosphere at about the sea-level"; for under a greater pressure water will not boil at that temperature, whilst under less pressure it boils at a lower temperature. In the usual statement of a law of causation, 'disturbing,' 'frustrating,' 'counteracting' circumstances (that is, negative conditions) are supposed to be absent; so that the strict statement of such a law, whether for a remote cause, or for an immediate cause (when only positive conditions are included), is that the agent or assemblage of conditions, *tends* to produce such an effect, other conditions being favourable, or in the absence of contrary forces.

(11) It is needless to repeat what has already been said of other fallacies that beset inductive proof; such as the neglect of a possible plurality of causes where the effect has been vaguely conceived; the extension of empirical laws beyond adjacent cases; the chief errors to which the estimate of analogies and probabilities, or the application of the principles of classification are liable; and the reliance upon direct Induction where the aid of Deduction may be obtained, or upon observation where experiment may be employed. As to formal fallacies that may be avoided by adhering to the rules of logical method, this may suffice.

§ 4. There remain many ways in which arguments fall short of a tolerable standard of proof, though they cannot be exhibited as definite breaches of logical principles. Logicians, therefore, might be excused from discussing them; but out of the abundance of their pity for human infirmity they usually describe and label the chief classes of these 'extra-logical fallacies,' and exhibit a few examples.

We may adopt Whately's remark, that a fallacy lies either (1) in the premises, or (2) in the conclusion, or (3) in the attempt to connect a conclusion with the premises.

(1) Now the premises of a sound argument must either be valid deductions, or valid inductions, or particular observations, or axioms. In an unsound argument, then, whose premises are supported by either deduction or induction, the evidence may be reduced to logical rules; and its failure is therefore a 'logical fallacy' such as we have already discussed. It follows that an extra-logical fallacy of the premises must lie in what cannot be reduced to rules of evidence, that is, in bad observations (§ 5), or sham axioms (§ 6).

(2) As to the conclusion, this can only be fallacious if some other conclusion has been substituted for that which was to have been proved (§ 7).

(3) Fallacies in the connection between premises and conclusion, if all the propositions are distinctly and explicitly stated, become manifest upon applying the rules of Logic. Fallacies, therefore, which are not thus manifest, and so are extra-logical, must depend upon some sort of slurring, confusion, or ambiguity of thought or speech (§ 8).

§ 5. Amongst Fallacies of Observation, Mill distinguishes (1) those of Non-observation, where either instances of the presence or absence of the phenomenon under investigation, or else some of the circumstances constituting it or attending upon it, though important to the induction, are overlooked. These errors are implied in the Formal Fallacies of Induction already treated of in § 3 (paragraphs (3) to (7)).

Mill's class (2) comprises fallacies of Malobservation. Malobservation may be due to obtuseness or slowness of perception; and it is one advantage of the physical sciences as means of education, that the training involved in studying them tends to cure these defects—at least, within their own range.

But the occasion of error upon which Mill most insists, is our proneness to substitute a hasty inference for a just representation of the fact before us; as when a yachtsman, eager for marvels, sees a line of porpoises and takes them for the sea-serpent. Every one knows what it is to mistake a stranger for a friend, a leaf for a sparrow, one word for another. The wonder is that we are not oftener wrong; considering how small a part present sensation has in perception, and how much of every object observed is supplied by a sort of automatic judgment. You see something brown, which your perceptive mechanism classes with the appearance of a cow at such a distance; and instantly all the other properties of a cow are supplied from the resources of former experience: but on getting nearer, it turns out to be a log of wood. It is some protection against such errors to know that we are subject to them; and the Logician fulfils his duty in warning us accordingly. But the matter belongs essentially to Psychology; and whoever wishes to pursue it will find a thorough explanation in Prof. Sully's volume on *Illusions*.

Another error is the accumulation of useless, irrelevant observations, from which no proof of the point at issue can be derived. It has been said that an important part of an inductive inquirer's equipment consists in knowing what to observe. The study of any science educates this faculty by showing us what observations have been effective in similar cases; but something depends upon genius. Observation is generally guided by hypotheses: he makes the right observations who can frame the right hypotheses; whilst another overlooks things, or sees them all awry, because he is confused and perverted by wishes, prejudices or other false preconceptions; and still another gropes about blindly, noting this and docketing that to no purpose, because he has no hypothesis, or one so vague and ill-conceived that it sheds no light upon his path.

§ 6. The second kind of extra-logical Fallacy lying in the premises, consists in offering as evidence some assertion entirely baseless or nugatory, but expressed in such a way as to seem like a general truth capable of subsuming the proposition in dispute: it is generally known as *petitio principii*, or begging the question. The question may be begged in three ways:

(1) There are what Mill calls Fallacies *a priori*, mere assertions, pretending to be self-evident, and often sincerely accepted as such by the author and some infatuated disciples, but in which the cool spectator sees either no sense at all, or palpable falsity. These sham axioms are numerous; and probably every one is familiar with the following examples: That circular motion is the most perfect; That every body strives toward its natural place; That like cures like; That every bane has its antidote; That what is true of our conceptions is true of Nature; That pleasure is nothing but relief from pain; That the good, the beautiful and the true are the same thing; That, in trade, whatever is somewhere gained is somewhere lost; That only in agriculture does nature assist man; That a man may do what he will with his own; That some men are naturally born to rule and others to obey. Some of these doctrines are specious enough; whilst, as to others, how they could ever have been entertained arouses a wonder that can only be allayed by a lengthy historical and psychological disquisition.

(2) Verbal propositions offered as proof of some matter of fact. These have, indeed, one attribute of axioms; they are self-evident to any one who knows the language; but as they only dissect the meaning of words, nothing but the meaning of words can be inferred from them. If anything further is arrived at, it must be by the help of real propositions. How common is such an argument as this: 'Lying is wrong, because it is vicious'—the implied major premise being that 'what is vicious is wrong.' All three propositions are verbal, and we merely learn from them that lying is *called* vicious and wrong; and to make that knowledge deterrent, it must be supplemented by a further premise, that 'whatever is called wrong ought to be avoided.' This is a real proposition; but it is much more difficult to prove it than 'that lying ought to be avoided.' Still, such arguments, though bad Logic, often have a rhetorical force: to call lying not only wrong but vicious, may be dissuasive by accumulating associations of shame and ignominy.

Definitions, being the most important of verbal propositions (since they imply the possibility of as many other verbal propositions as there are defining attributes and combinations of them), need to be watched with especial care. If two disputants define the same word in different ways, with each of the different attributes included in their several definitions they may bring in a fresh set of real propositions as to the agency or normal connection of that attribute. Hence their conclusions about the things denoted by the word defined, diverge in all directions and to any extent. And it is generally felt that a man who is allowed to define his terms as he pleases, may prove anything to those who, through ignorance or inadvertence, grant that the things that those terms stand for have the attributes that figure in his definitions.

(3) *Circulus in demonstrando*, the pretence of giving a reason for an assertion, whilst in fact only repeating the assertion itself—generally in other words. In such cases the original proposition is, perhaps, really regarded as self-evident, but by force of habit a man says 'because'; and then, after vainly fumbling in his empty pocket for the coin of reason, the habit of symbolic thinking in words only, without reference to the facts, comes to his rescue, and he ends with a paraphrase of the same assertion. Thus a man may try to prove the necessity of Causation: 'Every event must have a cause; because an event is a change of phenomena, and this implies a transformation of something pre-existing; which can only have been possible, if there were forces in operation capable of transforming it.' Or, again: 'We ought not to go to war, because it is wrong to shed blood.' But, plainly, if war did not imply bloodshed, the unlawfulness of this could be nothing against war. The more serious any matter is, the more important it becomes either to reason thoroughly about it, or to content ourselves with wholesome assertions. How many 'arguments' are superfluous!

§ 7. The Fallacy of surreptitious conclusion (*ignoratio elenchi*), the mistaking or obscuring of the proposition really at issue, whilst proving something else instead. This may be done by substituting a particular proposition for an universal, or an universal for a particular. Thus, he who attacks the practice of giving in charity must not be content to show that it has, in this or that case, degraded the recipient; who may have been exceptionally weak. Or, again, to dissuade another from giving alms in a particular case, it is not enough to show that the general tendency of almsgiving is injurious; for, by taking pains in a particular case, the general tendency may often be counteracted.

Sometimes an argument establishing a wholly irrelevant conclusion is substituted for an *argumentum ad rem*. Macaulay complains of those apologists for Charles I. who try to defend him as a king, by urging that he was a good judge of paintings and indulgent to his wife.

To this class of Fallacies belongs the *argumentum ad hominem*, which consists in showing not that a certain proposition is true, but that Critias ought to accept it in consistency with his other opinions. Thus: 'In every parish the cost of education ought to be paid out of the rates: you, at least, have said that there can be no sound economy, unless local expenses are defrayed from local funds.' But whether this is a fallacy depends, as Whately observes, upon whether it is urged as actually proving the point at issue, or merely as convicting the opponent of inconsistency. In the latter case, the argument is quite fair: whatever such a conclusion may be worth.

Similarly with the *argumentum ad populum*: 'this measure is favourable to such or such a class; let them vote for it.' An appeal to private greed, however base, is not fallacious, as long as the interest of the class is not *fraudulently* substituted for the good of the nation. And much the same may be said for the *argumentum ad verecundiam*. When a question of morals is debated as a question of honour among thieves, there is no fallacy, if the moral issue is frankly repudiated. The argument from authority is often brought under this head: 'such is the opinion of Aristotle.' Although this does not establish the truth of any proposition, it may be fairly urged as a reason for not hastily adopting a contrary conclusion: that is, if the subject under discussion be one as to which Aristotle (or whoever the authority may be) had materials for forming a judgment.

A negative use of this fallacy is very common. Some general doctrine, such as Positivism, Transcendentalism, Utilitarianism, or Darwinism, is held in common by a group of men; who, however, all judge independently, and therefore are likely to differ in details. An opponent exhibits their differences of opinion, and thereupon pretends to have refuted the theory they agree in supporting. This is an *argumentum ad scholam*, and pushes too far the demand for consistency. In fact it recoils upon the sophist; for there is no sense in quoting men against one another, unless both (or all) are acknowledged to speak with the authority of learning and judgment, and therefore the general doctrine which they hold in common is the more confirmed.

This is an example of the paralogism of 'proving too much'; when a disputant is so eager to refute an opponent as to lay down, or imply, principles from which an easy inference destroys his own position. To appeal to a principle of greater sweep than the occasion requires may easily open the way to this pitfall: as if a man should urge that 'all men are liars,' as the premise of an argument designed to show that another's assertion is less credible than his own.

A common form of *ignoratio elenchi* is that which Whately called the 'fallacy of objections': namely, to lay stress upon all the considerations against any doctrine or proposal, without any attempt to weigh them against the considerations in its favour; amongst which should be reckoned all the considerations that tell against the alternative doctrines or proposals. Incontestable demonstration can rarely be expected even in science, outside of the Mathematics; and in practical affairs, as Butler says, 'probability is the very guide of life'; so that every conclusion depends upon the balance of evidence, and to allow weight to only a part of it is an evasion of the right issue.

§ 8. Fallacies in the connection of premises and conclusion, that cannot be detected by reducing the arguments to syllogistic form, must depend upon some juggling with language to disguise their incoherence. They may be generally described as Fallacies of Ambiguity, whether they turn upon the use of the same word in different senses, or upon ellipsis. Thus it may be argued that all works written in a classical language are classical, and that, therefore, the history of Philosophy by Diogenes Laertius, being written in Greek, is a classic. Such ambiguities are sometimes serious enough; sometimes are little better than jokes. For jokes, as Whately observes, are often fallacies; and considered as a propædeutic to the art of sophistry, punning deserves the ignominy that has overtaken it.

Fallacies of ellipsis usually go by learned names, as; (1) *a dicto secundum quid ad dictum simpliciter*. It has been argued that since, according to Ricardo, the value of goods depends solely upon the quantity of labour necessary to produce them, the labourers who are employed upon (say) cotton cloth ought to receive as wages the whole price derived from its sale, leaving nothing for interest upon capital. Ricardo, however, explained that by 'the quantity of labour necessary to produce goods' he meant not only what is immediately applied to them, but also the labour bestowed upon the implements and buildings with which the immediate labour is assisted. Now these buildings and implements are capital, the labour which produced them was paid for, and it was far enough from Ricardo's mind to suppose that the capital which assists present labour upon (say) cotton cloth has no claim to remuneration out of the price of it. In this argument, then, the word labour in the premise is used *secundum quid*, that is, with the suppressed qualification of including past as well as present labour; but in the conclusion labour is used *simpliciter* to mean present labour only.

(2) *A dicto secundum quid ad dictum secundum alterum quid*. It may be urged that, since the tax on tea is uniform, therefore all consumers contribute equally to the revenue for their enjoyment of it. But written out fairly this argument runs thus: Since tea is taxed uniformly *4d. per lb.*, all consumers pay equally for their enjoyment of it *whatever quantity they use*. These qualifications introduced, nobody can be deceived.

(3) *A dicto simpliciter ad dictum secundum quid*, also called *fallacia accidentis*. Thus: To take interest upon a loan is perfectly just, therefore, I do right to exact it from my own father in distress. The popular answer to this sort of blunder is that 'circumstances alter cases.' We commit this error in supposing that what is true of the average is likely to be true of each case; as if one should say: 'The offices are ready to insure my house against fire at a rate per annum which will leave them heavy losers unless it lasts a hundred years; so, as we are told not to take long views of life, I shall not insure.'

The Fallacy of Division and Composition consists in suggesting, or assuming, that what is true of things severally denoted by a term is true of them taken together. That every man is mortal is generally admitted, but we cannot infer that, therefore, the human race will become extinct. That the remote prospects of the race are tragic may be plausibly argued, but not from that premise.

Changing the Premises is a fallacy usually placed in this division; although, instead of disguising different meanings under similar words, it generally consists in using words or phrases ostensibly differing, as if they were equivalent: those addressed being expected to renounce their right to reduce the argument to strict forms of proof, as needless pedantry in dealing with an author so palpably straightforward. If an orator says—'Napoleon conquered Europe; in other words, he murdered five millions of his fellow creatures'—and is allowed to go on, he may infer from the latter of these propositions many things which the former of them would hardly have covered. This is a sort of hyperbole, and there is a corresponding meiosis, as: 'Mill *admits* that the Syllogism is useful'; when, in fact, that is Mill's *contention*. It may be supposed that, if a man be fool enough to be imposed upon by such transparent colours, it serves him right; but this harsh judgment will not be urged by any one who knows and considers the weaker brethren.

§ 9. The above classification of Fallacies is a rearrangement of the plans adopted by Whately and Mill. But Fallacies resemble other spontaneous natural growths in not submitting to precise and definite classification. The same blunders, looked at from different points of view, may seem to belong to different groups. Thus, the example given above to illustrate *fallacia accidentis*, 'that, since it is just to take interest, it is right to exact it from one's own father,' may also be regarded as *petitio principii*, if we consider the unconditional statement of the premise—'to take interest upon a loan is perfectly just'; for, surely, this is only conditionally true. Or, again, the first example given of simple ambiguity—'that whatever is written in a classical language is classical, *etc.*,' may, if we attend merely to the major premise, be treated as a bad generalisation, an undue extension of an inference, founded upon a simple enumeration of the first few Greek and Latin works that one happened to remember.

It must also be acknowledged that genuine wild fallacies, roaming the jungle of controversy, are not so easily detected or evaded as specimens seem to be when exhibited in a Logician's collection; where one surveys them without fear, like a child at a menagerie. To assume the succinct mode of statement that is most convenient for refutation, is not the natural habit of these things. But to give reality to his account of fallacies an author needs a large space, that he may quote no inconsiderable part of literature ancient and modern.

As to the means of avoiding fallacies, a general increase of sincerity and candour amongst mankind may be freely recommended. With more honesty there would be fewer bad arguments; but there is such a thing as well-meaning incapacity that gets unaffectedly fogged in converting A., and regards the refractoriness of O., as more than flesh and blood can endure. Mere indulgence in figurative language, again, is a besetting snare. "One of the fathers, in great severity called poesy *vinum dæmonum*," says Bacon: himself too fanciful for a philosopher. Surely, to use a simile for the discovery of truth is like studying beauty in the bowl of a spoon.

The study of the natural sciences trains and confirms the mind in a habit of good reasoning, which is the surest preservative against paralogism, as long as the terms in use are, like those of science, well defined; and where they are ill defined, so that it is necessary to guard against ambiguity, a thorough training in politics or metaphysics may be useful. Logic seems to me to serve, in some measure, both these purposes. The conduct of business, or experience, a sufficient time being granted, is indeed the best teacher, but also the most austere and expensive. In the seventeenth century some of the greatest philosophers wrote *de intellectus emendatione*; and if their successors have given over this very practical inquiry, the cause of its abandonment is not success and satiety but despair. Perhaps the right mind is not to be made by instruction, but can only be bred: a slow, haphazard process; and meanwhile the rogue of a sophist may count on a steady supply of dupes to amuse the tedium of many an age.

FINIS.

Printed in Great Britain
by Amazon.co.uk, Ltd.,
Marston Gate.